© 2024 by FAISAL JAMIL. All rights reserved.

Title: "Climate Change and Its Effects on Human Identity"

This book, along with its contents encompassing text, illustrations, images, diagrams, and other creative elements, is the exclusive property of FAISAL JAMIL and is safeguarded by copyright law.

FAISAL JAMIL asserts full ownership and retains all rights to this book. No part of this publication may be reproduced, distributed, or transmitted in any form or by any means, such as photocopying, recording, or electronic methods, without prior written consent from the copyright holder. Brief quotations in critical reviews and certain noncommercial uses permitted by copyright law are exceptions.

This copyright notice applies to all editions, formats, and translations of the book, whether in print, digital, or any other medium or technology existing now or developed in the future. Unauthorized use or infringement may result in legal action and pursuit of remedies under applicable copyright laws.

While efforts have been made to ensure accuracy and reliability, FAISAL JAMIL does not guarantee the completeness or suitability of the information. Readers are responsible for evaluating and using the content judiciously.

FAISAL JAMIL reserves the right to make changes, updates, or corrections to the book without prior notice. Inclusion of third-party materials or references does not imply endorsement or affiliation unless used under fair use principles or with proper permissions and attributions.

For permissions, inquiries, or requests regarding the book's use, please contact FAISAL JAMIL through official channels listed on their Amazon author page or provided email address.

This comprehensive copyright notice serves to protect FAISAL JAMIL'S intellectual property rights, maintain content control, and inform users about associated restrictions and permissions.

Warm regards,

FAISAL JAMIL

For Your Feedback and Reviews!

http://www.amazon.com/author/faisal.jamil

Email: faisaljamilauthor@gmail.com

About the author

Certainly! Faisal Jamil is a multifaceted individual with a diverse set of skills and experiences. With a strong foundation in computer knowledge since childhood, he has developed a deep understanding of technology that informs his work as a content writer. Faisal also possesses digital skills, which further enhance his abilities in various digital platforms and technologies.

Beyond his professional endeavors, Faisal Jamil has also excelled in the martial arts, particularly Shotokan Karate, where he achieved the prestigious rank of first Dan black belt. This achievement speaks to his dedication, discipline, and commitment to personal growth and mastery.

In his professional life, Faisal Jamil has carved out a successful career in sales management within the Fast Moving Consumer Goods (FMCG) sector. His roles in various FMCG companies have honed his skills in strategic planning, team leadership, and business development. Faisal's ability to drive sales and achieve targets has been instrumental in his career progression, showcasing his talent for identifying opportunities and delivering results.

Faisal Jamil is also deeply interested in business investment strategies, planning, and execution. His understanding of these areas has been key to his success in the business world, allowing him to make informed decisions and implement effective strategies. His ability to navigate the complexities of investment planning and execution has set him apart as a strategic thinker and a valuable asset in any business endeavor.

Overall, Faisal Jamil is a dynamic individual who combines his passion for technology, martial arts, sales management, digital skills, and business investment strategies to achieve success in diverse fields. His journey is a testament to his versatility, resilience, and continuous pursuit of excellence.

Yours Sincerely

FAISAL JAMIL

For Your Feedback and Reviews!

https://www.amazon.com/author/faisal.jamil

Email: faisaljamilauthor@gmail.com

CLIMATE CHANGE
AND ITS EFFECTS ON
HUMAN IDENTITY

Table of Content

Preface --- **11**

Introduction --- **14**

Chapter 1: Introduction to Climate Change and Human Identity --- **17**

Defining Climate Change: Science and Perception

Human Identity: Evolution and Adaptation

Intersecting Realities: Climate and Identity

Chapter 2: Historical Perspectives on Climate and Human Societies --- **22**

Ancient Climate Events and Cultural Shifts

Climate's Role in Civilizational Rise and Fall

Lessons from History: Adaptation and Resilience

Chapter 3: Modern Climate Change: Causes and Consequences -- **27**

The Science of Anthropogenic Climate Change

Environmental Impacts and Feedback Loops

Societal Implications: Health, Economy, and Politics

Chapter 4: Identity in Crisis: Psychological Effects of Climate Change --- **34**

Eco-Anxiety: Mental Health in a Changing Climate

Loss of Place: Displacement and Identity Crisis

Finding Meaning in Uncertainty: Psychological Resilience

Chapter 5: Climate Change and Cultural Identity ---------41

Cultural Heritage at Risk: Threatened Traditions

Indigenous Knowledge and Climate Resilience

Globalization and Cultural Homogenization

Chapter 6: Climate Refugees: Displacement and Identity Transformation --48

The Rise of Climate-Induced Migration

Identity in Transit: Redefining Home and Belonging

Policy and Human Rights: Protecting the Displaced

Chapter 7: Economic Inequality and Climate Change ----55

The Rich and the Vulnerable: Disparate Impacts

Climate Justice: Bridging Economic Divides

Redefining Success: Sustainable Economies

Chapter 8: Gender and Climate Change ---------------------61

Gendered Impacts: Women and Climate Vulnerability

Empowering Women: Agents of Change

Intersectionality: Understanding Diverse Experiences

Chapter 9: Youth Identity in the Age of Climate Change-68

The Climate Generation: Youth Activism

Future Prospects: Education and Career Shifts

Building a Sustainable Identity: Youth Leadership

Chapter 10: Technological Identity in a Warming World-74

Innovation and Adaptation: Technological Solutions

Digital Activism: Online Movements and Climate Awareness

Ethical Considerations: Technology and Environmental Impact

Chapter 11: Climate Change and National Identity -------82

National Narratives: Climate Change in Policy and Media

Sovereignty and Responsibility: International Dynamics

National Pride vs. Global Responsibility: Balancing Acts

Chapter 12: Religion, Spirituality, and Climate Change--89

Faith-Based Responses: Religious Environmentalism

Spiritual Identity: Connecting with Nature

Interfaith Dialogue: Unified Climate Action

Chapter 13: Art and Expression in the Climate Crisis ----96

Climate Change in Literature and Film

Visual Arts: Documenting and Interpreting Change

Performing Arts: Advocacy through Creativity

Chapter 14: Climate Change and Urban Identity --------104

Urbanization and Environmental Footprints

Resilient Cities: Redesigning for Sustainability

Community Identity: Urban Green Spaces and Social Cohesion

Chapter 15: Rural and Agricultural Identities in Transition ---112

Changing Landscapes: Agriculture and Climate

Farmers and Fishers: Adapting Traditional Livelihoods

Rural Community Resilience: Collective Adaptation

Chapter 16: The Role of Education in Shaping Climate Identity ---121

Curriculum Changes: Integrating Climate Education

Critical Thinking: Empowering Informed Citizens

Lifelong Learning: Community and Global Initiatives

Chapter 17: Policy, Governance, and Identity -------------129

Climate Policy: Local, National, and Global Frameworks

Governance and Trust: Public Perception and Engagement

Identity Politics: Advocacy and Representation

Chapter 18: Ethical and Philosophical Dimensions of Climate Change ---139

Moral Imperatives: Responsibility and Stewardship

Philosophical Debates: Human-Nature Relationships

Ethical Frameworks: Guiding Climate Action

Chapter 19: Imagining the Future: Scenarios and Identity — 148

Possible Futures: Optimistic, Pessimistic, and Realistic

Adapting Identities: Personal and Collective Narratives

Building Hope: Visionary Leadership and Innovation

Chapter 20: Conclusion: Integrating Knowledge and Action — 158

Synthesis: Understanding the Interconnections

Empowerment: From Awareness to Action

Future Directions: Continuing the Journey

Epilogue: A Call to Action — 166

Individual Responsibility: Everyday Choices

Collective Effort: Community and Global Initiatives

Sustaining Change: A Unified Future

Thank You for Reading — 173

Preface

Climate change is one of the most pressing issues of our time, affecting every aspect of our lives in ways that are both profound and pervasive. It challenges our physical environments, economic systems, social structures, and, perhaps most importantly, our very sense of identity. The ways in which we perceive ourselves, our communities, and our place in the world are being reshaped by the realities of a changing climate.

This book, "Climate Change and Its Effects on Human Identity," seeks to explore these complex and often overlooked dimensions of climate change. While much attention has been given to the scientific and environmental aspects of this global crisis, less focus has been placed on how it intersects with human identity. Our identities are not static; they evolve in response to the world around us. As climate change alters landscapes, economies, and social fabrics, it also transforms our identities, influencing how we understand ourselves and relate to others.

In the pages that follow, we will journey through various facets of this transformation. We will examine historical perspectives to understand how past climate events have shaped human societies and cultures. We will delve into the psychological impacts of climate change, exploring issues such as eco-anxiety and the loss of place. The book will also highlight the resilience and adaptability of human identity, showcasing how communities are responding to and mitigating the effects of climate change.

We will look at the role of cultural identity, particularly the experiences of indigenous peoples and the threats to cultural heritage. The rise of climate refugees and the redefinition of home and belonging will be discussed, as well as the economic and gender inequalities exacerbated by climate change. The voices and activism of youth, the influence of technology, and the narratives of national identity will all be explored in the context of a warming world.

Moreover, we will consider the contributions of art, religion, and spirituality in shaping climate identity and fostering a sense of connection to the natural world. Urban and rural identities, the role of education, and the ethical and philosophical dimensions of climate change will also be addressed, providing a comprehensive overview of how deeply intertwined climate change is with every aspect of human existence.

This book is not just an academic exploration but a call to action. It emphasizes the power of individual responsibility and the importance of collective effort. By understanding the interconnections between climate change and human identity, we can better appreciate the urgency of the situation and the need for sustained, unified action.

As you read through these chapters, I hope you will find not only knowledge and insight but also inspiration and hope. The challenges are immense, but so too is our capacity for resilience, innovation, and change. Together, we can shape a future where both humanity and the planet thrive.

Thank you for embarking on this journey with me.

FAISAL JAMIL

Introduction

Climate change is not a distant threat but a present reality that impacts every corner of our world. Its effects extend beyond melting ice caps and rising sea levels to touch the very core of our existence: our identity. How we see ourselves, our communities, and our connection to the planet is being reshaped by this profound crisis. This book, "Climate Change and Its Effects on Human Identity," seeks to unravel the complex interplay between a changing climate and the evolution of human identity.

Defining Climate Change: Science and Perception

Climate change refers to long-term alterations in temperature, precipitation patterns, and other atmospheric conditions on Earth. While natural processes have always driven climate variability, the current trend of rapid warming is predominantly due to human activities, such as the burning of fossil fuels, deforestation, and industrial processes. Scientific consensus underscores the urgency of addressing these changes to mitigate their impacts on ecosystems and human societies.

However, climate change is not merely a scientific concept; it is also a social and psychological phenomenon. People's perceptions of climate change are shaped by their cultural backgrounds, personal experiences, and media representations. Understanding these perceptions is crucial for fostering meaningful dialogue and effective action.

Human Identity: Evolution and Adaptation

Human identity is a multifaceted construct encompassing individual and collective senses of self. It is influenced by cultural heritage, personal experiences, social interactions, and environmental contexts. Throughout history, human identities have evolved in response to various challenges and changes, including those posed by the natural world.

Adaptation is a key aspect of human identity. As environments change, so do the ways in which people live, work, and relate to one another. This adaptability is both a strength and a source of anxiety as the pace and scale of contemporary climate change present unprecedented challenges. How we adapt to these changes—physically, emotionally, and culturally—will define the future of human identity.

Intersecting Realities: Climate and Identity

The intersection of climate change and human identity reveals a tapestry of interconnected impacts and responses. For many, the changing climate threatens cultural heritage, disrupts traditional livelihoods, and forces migration. Indigenous communities, in particular, face profound threats to their cultural and spiritual connections to the land. At the same time, new forms of identity and solidarity are emerging as people unite in the face of a common challenge.

This book explores these intersecting realities through a series of focused chapters. We begin with historical perspectives, examining how past climate events have shaped civilizations. We then move to contemporary issues, including the psychological effects of eco-anxiety, the cultural impacts on traditions and heritage, and the experiences of climate refugees. The role of economic inequality, gender dynamics, youth activism, technological advancements, and national narratives in shaping climate identity will also be discussed.

Furthermore, we delve into the contributions of religion, spirituality, and the arts in fostering a sense of connection and resilience. Urban and rural identities, educational initiatives, policy frameworks, and ethical considerations are all part of this intricate web. Finally, we look toward the future, envisioning possible scenarios and the potential for transformative change.

Through this exploration, we aim to provide a comprehensive understanding of how climate change is not only an environmental crisis but also a profound social and cultural phenomenon. By recognizing the deep connections between climate and identity, we can better appreciate the stakes involved and the need for collective action. This book is a call to understand, adapt, and act—embracing our evolving identities in the face of a changing world.

Chapter 1
Introduction to
Climate Change and
Human Identity

Defining Climate Change: Science and Perception

The Science of Climate Change

Climate change refers to significant and lasting changes in the statistical distribution of weather patterns over periods ranging from decades to millions of years. It can be caused by natural factors such as volcanic eruptions, variations in solar radiation, and plate tectonics. However, the current phase of climate change, which has become particularly noticeable since the late 19th century, is primarily driven by human activities. The burning of fossil fuels, deforestation, and industrial processes have led to an increase in greenhouse gases (GHGs) like carbon dioxide (CO_2), methane (CH_4), and nitrous oxide (N_2O) in the Earth's atmosphere. These gases trap heat, leading to a warming effect known as the greenhouse effect.

Perceptions and Misconceptions

Despite the overwhelming scientific consensus on the reality and causes of climate change, public perception varies widely. Factors influencing perception include media representation, political ideologies, cultural beliefs, and personal experiences. Some individuals accept the scientific findings and are concerned about the impacts of climate change, while others are skeptical or outright deny its existence or human causation. Misconceptions about climate change are often fueled by misinformation and a lack of understanding of the scientific method and data interpretation.

Bridging the Gap

To effectively address climate change, it is crucial to bridge the gap between scientific knowledge and public perception. This involves improving science communication, enhancing climate education, and fostering a culture of critical thinking and skepticism that evaluates evidence objectively. By doing so, society can form a more unified and informed approach to mitigating and adapting to climate change.

Human Identity: Evolution and Adaptation

Evolution of Human Identity

Human identity is a complex and dynamic construct shaped by various factors, including biology, culture, environment, and personal experiences. Historically, human identity has evolved through stages marked by significant changes in living conditions, social structures, and technological advancements. From the hunter-gatherer societies of the prehistoric era to the agrarian and industrial revolutions, each phase of human history has left an indelible mark on our collective and individual identities.

Adaptation to Environmental Changes

Throughout history, humans have adapted to environmental changes in numerous ways. Early humans developed tools and clothing to survive in diverse climates. Agricultural practices evolved in response to changing weather patterns, and entire civilizations relocated or adapted their ways of life due to environmental shifts. These adaptations have not only ensured survival but also influenced cultural practices, societal norms, and collective identities.

The Role of Resilience

Resilience, the capacity to recover from difficulties and adapt to change, is a critical component of human identity. In the context of climate change, resilience involves both psychological and practical dimensions. On a psychological level, it includes the ability to cope with climate-related stress and anxiety. Practically, it encompasses strategies and actions taken to mitigate and adapt to the impacts of climate change. Building resilience is essential for maintaining a stable and coherent identity in the face of ongoing environmental challenges.

Intersecting Realities: Climate and Identity

The Interconnection of Climate and Identity

Climate and human identity are deeply interconnected. The environment influences culture, social structures, and individual behaviors, which in turn shape identities. Climate change, by altering the environment, has the potential to profoundly impact how people see themselves and their place in the world. This intersection is particularly evident in communities directly dependent on natural resources and ecosystems, where changes in climate can disrupt traditional ways of life and cultural practices.

Identity in the Face of Change

As climate change accelerates, individuals and communities are faced with the challenge of maintaining their identities amidst changing circumstances. This can involve redefining what it means to belong to a particular place, community, or cultural group. For instance, indigenous communities whose identities are closely tied to their ancestral lands may experience profound identity shifts as those lands are altered or lost. Similarly, individuals who identify with specific climate-related occupations, such as farming or fishing, may need to adapt their identities in response to changing environmental conditions.

Towards a Sustainable Identity

Building a sustainable identity involves integrating awareness of climate change into the fabric of personal and collective identity. This means fostering a sense of responsibility towards the environment and future generations. It also involves embracing adaptive practices and resilience as core aspects of identity. By doing so, individuals and communities can navigate the challenges posed by climate change while maintaining a coherent and meaningful sense of self.

This chapter sets the stage for a deeper exploration of the complex relationship between climate change and human identity, providing a foundational understanding of the scientific, psychological, and cultural dimensions involved.

Chapter 2
Historical Perspectives on Climate and Human Societies

Ancient Climate Events and Cultural Shifts

Climate Events in Prehistoric Times

Throughout prehistory, climatic changes have played a crucial role in shaping human development and migration. The Ice Ages, characterized by long periods of glacial expansion and retreat, significantly influenced human evolution and settlement patterns. For instance, the harsh conditions of the Last Glacial Maximum, which peaked around 20,000 years ago, pushed early humans to adapt to colder environments, leading to technological advancements such as the development of clothing and shelter.

The Holocene and the Agricultural Revolution

The end of the last Ice Age around 11,700 years ago marked the beginning of the Holocene epoch, a period of relatively stable and warmer climate. This stability facilitated the transition from nomadic hunter-gatherer societies to settled agricultural communities. The Neolithic Revolution, which began around 10,000 years ago in the Fertile Crescent, was largely driven by the favorable climatic conditions that allowed for the domestication of plants and animals. This shift fundamentally transformed human societies, leading to the development of permanent settlements, social stratification, and complex civilizations.

Cultural Shifts in Response to Climate Variability

Various ancient cultures experienced significant shifts in response to climatic variability. The decline of the Akkadian Empire around 2200 BCE, for instance, is believed to be associated with a severe drought that disrupted agriculture and led to social unrest. Similarly, the collapse of the Classic Maya civilization around the 9th century CE has been linked to prolonged periods of drought, which undermined the region's agricultural base and contributed to political instability. These examples highlight the profound impact of climate events on cultural evolution and societal change.

Climate's Role in Civilizational Rise and Fall

The Rise of Civilizations

Several early civilizations emerged in regions with favorable climatic conditions that supported agriculture and population growth. The Nile Valley in Egypt, the Tigris and Euphrates river valleys in Mesopotamia, and the Indus Valley in South Asia all benefitted from predictable flooding patterns and fertile soils, which enabled the development of sophisticated agricultural systems. These conditions supported the growth of large, complex societies with advanced technologies, governance structures, and cultural achievements.

The Fall of Civilizations

Conversely, climatic changes have also played a significant role in the decline and fall of civilizations. The collapse of the Harappan civilization in the Indus Valley around 1900 BCE has been attributed to changes in monsoon patterns that led to a decline in agricultural productivity. Similarly, the fall of the Western Roman Empire in the 5th century CE coincided with a period of climatic instability known as the Late Antique Little Ice Age, which contributed to crop failures, economic decline, and invasions by migrating tribes.

The Medieval Climate Anomaly and the Little Ice Age

The Medieval Climate Anomaly (950-1250 CE) and the subsequent Little Ice Age (1300-1850 CE) are examples of climatic periods that had significant impacts on human societies. The Medieval Climate Anomaly was characterized by warmer temperatures in some regions, which facilitated agricultural expansion and population growth in Europe. In contrast, the Little Ice Age brought colder temperatures, leading to shorter growing seasons, crop failures, and famines. These climatic fluctuations influenced historical events such as the Viking expansion, the Black Death, and the social and economic turmoil of the 17th century.

Lessons from History: Adaptation and Resilience

Historical Adaptations to Climate Change

Historical societies have employed various strategies to adapt to climatic changes. The ancient Egyptians developed sophisticated irrigation systems to manage the Nile's flooding and ensure reliable agricultural production. Similarly, the Inca civilization in the Andes constructed extensive terracing and irrigation networks to maximize agricultural output in a challenging mountainous environment. These adaptations highlight the ingenuity and resilience of human societies in the face of environmental challenges.

Social and Political Responses

Climate-induced crises have often prompted significant social and political changes. The Dust Bowl of the 1930s, caused by severe drought and poor agricultural practices in the United States, led to widespread displacement and economic hardship. In response, the U.S. government implemented policies such as the New Deal, which included measures to restore the environment and provide social support. This example illustrates how societies can mobilize resources and implement policy changes to address the impacts of climatic events.

Building Resilience for the Future

The historical record provides valuable lessons for contemporary efforts to build resilience to climate change. Key strategies include diversifying livelihoods, investing in infrastructure that can withstand climatic extremes, and fostering social cohesion and cooperation. Additionally, learning from past successes and failures can inform current climate policies and practices, enabling societies to better anticipate, prepare for, and respond to future climatic challenges.

This chapter highlights the deep historical interconnections between climate events and human societies, demonstrating how climatic changes have shaped cultural evolution, the rise and fall of civilizations, and the development of adaptive strategies.

These insights provide a foundation for understanding the ongoing and future impacts of climate change on human identity and social structures.

Chapter 3
Modern Climate Change Causes and Consequences

The Science of Anthropogenic Climate Change

Greenhouse Gases and Their Sources

Anthropogenic (human-caused) climate change is primarily driven by the increase in greenhouse gases (GHGs) in the atmosphere. The main GHGs contributing to climate change include carbon dioxide (CO_2), methane (CH_4), nitrous oxide (N_2O), and fluorinated gases. CO_2 is largely emitted from the burning of fossil fuels (coal, oil, and natural gas) for energy and transportation, as well as deforestation and land-use changes. Methane emissions arise from agriculture (especially livestock production), landfills, and natural gas extraction. Nitrous oxide is released from agricultural activities, including the use of synthetic fertilizers, and various industrial processes. Fluorinated gases, though less abundant, are potent GHGs used in refrigeration, air conditioning, and other industrial applications.

The Greenhouse Effect

The greenhouse effect is a natural process where certain gases in the Earth's atmosphere trap heat from the sun, keeping the planet warm enough to sustain life. However, the excessive increase in GHGs due to human activities has enhanced this effect, leading to a rise in global temperatures. This enhanced greenhouse effect results in more heat being trapped in the atmosphere, causing the Earth's average surface temperature to increase, a phenomenon known as global warming.

Evidence of Anthropogenic Climate Change

Multiple lines of evidence support the reality of anthropogenic climate change. These include:

1: Temperature Records: Instrumental temperature records show a clear increase in global average temperatures over the past century, with the most significant warming occurring in the last few

2: Ice Core Data: Ice cores extracted from glaciers and ice sheets reveal historical GHG concentrations and temperature variations, showing a strong correlation between GHG levels and global temperatures.

3: Ocean Heat Content: The world's oceans have absorbed much of the excess heat, leading to rising sea surface temperatures and thermal expansion.

4: Sea Level Rise: Melting ice sheets and glaciers, combined with thermal expansion of seawater, have contributed to rising sea levels.

5: Changes in Weather Patterns: Increasing frequency and intensity of extreme weather events, such as heatwaves, droughts, and heavy precipitation, align with climate change predictions.

Environmental Impacts and Feedback Loops

Environmental Impacts

The environmental consequences of climate change are far-reaching and multifaceted:

1: Melting Ice and Rising Sea Levels: Polar ice caps and glaciers are melting at unprecedented rates, contributing to sea level rise. This threatens coastal communities, ecosystems, and freshwater

2: Ocean Acidification: Increased CO2 levels lead to higher concentrations of carbonic acid in the oceans, negatively impacting marine life, particularly organisms with calcium carbonate shells or skeletons.

3: Ecosystem Disruption: Shifts in temperature and precipitation patterns disrupt ecosystems, affecting biodiversity, species distribution, and the timing of biological events (phenology).

4: Extreme Weather Events: Climate change exacerbates the frequency and intensity of extreme weather events, including hurricanes, heatwaves, floods, and droughts, leading to significant environmental and economic damage.

Feedback Loops

Climate feedback loops can either amplify or dampen the effects of climate change:

1: Positive Feedback Loops:

Ice-Albedo Feedback: As ice melts, it exposes darker surfaces (e.g., ocean or land), which absorb more solar radiation, leading to further warming and more ice melt.

Permafrost Thaw: Thawing permafrost releases stored methane, a potent GHG, into the atmosphere, accelerating warming.

Forest Dieback: Increased temperatures and drought stress can lead to forest dieback, reducing carbon sequestration and releasing more CO2 into the atmosphere.

2: Negative Feedback Loops:

Increased Vegetation Growth: Higher CO2 levels can stimulate plant growth, which may enhance carbon sequestration (though this effect is limited by other factors such as nutrient availability and water stress).

Cloud Cover Changes: Increased evaporation and cloud formation can reflect more sunlight, potentially cooling the Earth (though this effect is complex and varies regionally).

Understanding these feedback mechanisms is crucial for predicting and mitigating the impacts of climate change.

Societal Implications: Health, Economy, and Politics

Health Implications

Climate change poses significant risks to public health:

1: Heat-Related Illnesses: Increased frequency and severity of heatwaves lead to higher rates of heat exhaustion, heatstroke, and related mortality, particularly among vulnerable populations.

2: Vector-Borne Diseases: Changes in temperature and precipitation can expand the range of vector-borne diseases, such as malaria, dengue fever, and Lyme disease, as vectors like mosquitoes and ticks find new suitable habitats.

3: Air Quality: Higher temperatures and altered weather patterns can worsen air pollution, increasing the prevalence of respiratory and cardiovascular diseases.

4: Food and Water Security: Climate change affects crop yields and water availability, leading to malnutrition and foodborne illnesses. Extreme weather events can also disrupt water supply and sanitation infrastructure.

Economic Implications

The economic impacts of climate change are profound and multifaceted:

1: Damage to Infrastructure: Extreme weather events cause significant damage to infrastructure, including roads, bridges, buildings, and power grids, resulting in substantial repair and replacement costs.

2: Agricultural Productivity: Changes in temperature and precipitation patterns affect agricultural productivity, leading to crop failures, reduced yields, and higher food prices. This can exacerbate poverty and food insecurity.

3: Insurance Costs: The increasing frequency and severity of natural disasters drive up insurance premiums and lead to higher costs for disaster recovery and rebuilding.

4: Migration and Displacement: Climate-induced displacement and migration strain resources and infrastructure in receiving areas, leading to economic and social challenges.

Political Implications

Climate change also has significant political ramifications:

1: International Relations: Climate change is a global issue requiring international cooperation. Disparities in contributions to GHG emissions and vulnerabilities to climate impacts can lead to tensions between countries.

2: Policy and Regulation: Governments must balance economic growth with environmental protection. Implementing effective climate policies, such as carbon pricing, renewable energy incentives, and emission reduction targets, is essential but often politically challenging.

3: Social Equity: Climate change disproportionately affects marginalized and vulnerable communities. Ensuring climate justice and equitable distribution of resources and support is a critical political issue.

4: National Security: Climate change can exacerbate resource conflicts, increase the risk of political instability, and pose threats to national security by destabilizing regions and contributing to mass migration.

This chapter provides a comprehensive overview of the causes and consequences of modern climate change, emphasizing the scientific, environmental, and societal dimensions. Understanding these aspects is crucial for developing effective strategies to mitigate and adapt to the impacts of climate change, ensuring a sustainable and resilient future.

Chapter 4
Identity in Crisis
Psychological Effects of Climate Change

Eco-Anxiety: Mental Health in a Changing Climate

Understanding Eco-Anxiety

Eco-anxiety refers to the chronic fear of environmental doom and the stress associated with the awareness of climate change and its potential impacts. It is a form of psychological distress that arises from concern over the planet's future and the potential threats to human life and the natural world. This anxiety is compounded by the sense of helplessness many feel in the face of such a large, complex, and often seemingly uncontrollable issue.

Manifestations of Eco-Anxiety

Eco-anxiety can manifest in various ways, including:

Generalized Anxiety and Stress: Persistent worry about the future of the planet and the wellbeing of future generations.

Depression: Feelings of sadness, hopelessness, and helplessness regarding the state of the environment and the perceived inability to make a meaningful impact.

Obsessive Thoughts: Fixation on environmental degradation, climate disasters, and the ongoing loss of biodiversity.

Somatic Symptoms: Physical symptoms such as headaches, muscle tension, and fatigue resulting from chronic stress and anxiety.

Demographic Variations

Eco-anxiety affects different demographic groups in various ways. Young people, who are more likely to witness the long-term impacts of climate change, often report higher levels of eco-anxiety. Indigenous communities and people living in areas most vulnerable to climate impacts, such as coastal regions and low-income countries, also experience heightened anxiety due to their direct exposure to environmental changes and their often limited resources for coping and adaptation.

Addressing Eco-Anxiety

Managing eco-anxiety involves a combination of personal, community, and societal efforts:

Personal Strategies: Mindfulness practices, therapy, and engaging in environmental activism can help individuals cope with eco-anxiety.

Community Support: Building supportive communities that validate and address eco-anxiety can reduce feelings of isolation and helplessness.

Societal Action: Governments and organizations can help by acknowledging eco-anxiety as a legitimate concern and implementing policies that address the root causes of climate change.

Loss of Place: Displacement and Identity Crisis

The Concept of "Place" and Identity

The concept of "place" is integral to human identity. It encompasses the physical, emotional, and cultural connections individuals have with their surroundings. A sense of place provides stability, belonging, and a framework for personal and collective identity. When this connection is disrupted, it can lead to a profound identity crisis.

Climate-Induced Displacement

Climate change is causing an increasing number of people to be displaced from their homes due to rising sea levels, extreme weather events, and changing environmental conditions that make traditional ways of life unsustainable. This displacement disrupts the sense of place and can lead to significant psychological and social challenges:

Loss of Home: The physical loss of a home and community can be deeply traumatic, leading to feelings of grief, loss, and disorientation.

Cultural Displacement: Displacement can sever ties to cultural practices, traditions, and community networks, exacerbating the sense of identity loss.

Economic Impact: The economic instability resulting from displacement can further compound psychological distress and hinder the ability to rebuild a sense of place and identity.

Case Studies

Pacific Island Nations: Rising sea levels threaten to submerge entire islands, forcing residents to relocate and abandon their ancestral lands. This leads to a loss of cultural heritage and a profound sense of identity disruption.

Indigenous Communities in the Arctic: Thawing permafrost and changing ecosystems impact traditional hunting and fishing practices, challenging the cultural identity of Indigenous peoples who have relied on these practices for generations.

Coping with Displacement

Addressing the psychological impacts of displacement involves:

Creating New Connections: Helping displaced individuals establish new social networks and communities can mitigate feelings of isolation and loss.

Preserving Cultural Identity: Efforts to preserve and adapt cultural practices in new locations can help maintain a sense of continuity and belonging.

Providing Mental Health Support: Access to counseling and mental health services is crucial for individuals coping with the trauma of displacement.

Finding Meaning in Uncertainty: Psychological Resilience

The Importance of Psychological Resilience

Psychological resilience refers to the ability to adapt to and recover from stress and adversity. In the context of climate change, resilience involves maintaining a sense of purpose and meaning despite the uncertainties and challenges posed by a changing environment. Resilient individuals and communities are better equipped to navigate the psychological impacts of climate change and find constructive ways to cope and adapt.

Building Resilience Through Community

Strong, supportive communities play a vital role in fostering resilience:

Collective Action: Participating in community-based environmental initiatives can provide a sense of agency and empowerment, countering feelings of helplessness.

Social Support Networks: Building and maintaining strong social connections offer emotional support and practical assistance during times of stress and uncertainty.

Shared Narratives: Developing and sharing stories of resilience and adaptation can inspire hope and collective identity.

Personal Strategies for Resilience

Individuals can cultivate resilience through various personal practices:

Mindfulness and Self-Care: Engaging in mindfulness practices, physical activity, and self-care routines can help manage stress and maintain mental wellbeing.

Cognitive Flexibility: Developing the ability to adapt one's thinking and embrace change can enhance resilience in the face of environmental uncertainties.

Purpose and Meaning: Finding personal meaning and purpose, whether through environmental activism, community engagement, or other pursuits, can provide motivation and a sense of direction.

Role of Education and Awareness

Education plays a crucial role in building resilience by:

Increasing Awareness: Understanding the science and impacts of climate change can empower individuals to take informed actions and make meaningful contributions to mitigation and adaptation efforts.

Promoting Adaptive Skills: Teaching skills such as critical thinking, problem-solving, and adaptive management can prepare individuals to navigate the complexities of a changing climate.

Fostering Hope and Agency: Education that highlights success stories, solutions, and pathways to positive change can inspire hope and a sense of agency.

This chapter delves into the profound psychological effects of climate change on human identity, highlighting the importance of addressing eco-anxiety, managing the impacts of displacement, and fostering psychological resilience. By understanding and addressing these psychological dimensions, individuals and communities can better navigate the challenges of a changing climate and maintain a sense of purpose and identity.

Chapter 5
Climate Change and Cultural Identity

Cultural Heritage at Risk: Threatened Traditions

The Impact of Climate Change on Cultural Heritage

Cultural heritage encompasses the traditions, practices, artifacts, and sites that define a community's identity and historical continuity. Climate change poses significant threats to both tangible and intangible cultural heritage, affecting everything from historical buildings and sacred sites to traditional practices and oral histories.

Tangible Cultural Heritage

Rising sea levels, increased flooding, and extreme weather events threaten historical and cultural landmarks around the world:

Coastal Sites: Coastal cities and ancient sites, such as Venice, Italy, and the ancient city of Alexandria in Egypt, face the risk of inundation and erosion due to rising sea levels.

Archaeological Sites: Sites like the ruins of ancient civilizations in low-lying areas, such as the Mesopotamian cities in Iraq, are at risk from flooding and changing water tables.

Historic Buildings: Extreme weather events, such as hurricanes and typhoons, can cause irreparable damage to historic structures. For example, hurricanes have repeatedly threatened the preservation of colonial-era buildings in the Caribbean.

Intangible Cultural Heritage

Climate change also threatens intangible cultural heritage, including languages, music, rituals, and traditional knowledge:

Seasonal Festivals: Many cultural festivals and rituals are tied to specific seasonal events, such as harvests or solstices. Changes in climate can disrupt these natural cycles, affecting the timing and relevance of these traditions.

Traditional Livelihoods: Practices like traditional farming, fishing, and herding are deeply intertwined with cultural identities. Changing climate conditions can render these practices untenable, leading to the loss of associated knowledge and traditions.

Oral Histories and Stories: As communities are displaced or fragmented by climate change, the transmission of oral histories and stories that preserve cultural identity can be disrupted.

Efforts to Preserve Cultural Heritage

Efforts to preserve cultural heritage in the face of climate change include:

Documentation and Digitization: Recording and digitizing cultural artifacts, oral histories, and traditions can help preserve them for future generations.

Adaptive Use: Modifying and adapting cultural practices to new environmental realities can help keep traditions alive.

International Cooperation: Global initiatives, such as UNESCO's efforts to protect World Heritage Sites, play a crucial role in preserving cultural heritage threatened by climate change.

Indigenous Knowledge and Climate Resilience

The Value of Indigenous Knowledge

Indigenous communities around the world possess extensive knowledge about their local environments, developed through centuries of interaction with their surroundings. This knowledge, often referred to as traditional ecological knowledge (TEK), includes insights into weather patterns, biodiversity, sustainable resource management, and adaptation strategies.

Examples of Indigenous Knowledge

Agricultural Practices: Indigenous agricultural practices, such as the Andean terraces and irrigation systems used by the Inca, are designed to maximize productivity and resilience in challenging environments. These systems can offer valuable lessons for modern sustainable agriculture.

Water Management: Indigenous water management techniques, such as the qanats of Iran and the acequias of the American Southwest, demonstrate efficient water use and conservation practices in arid regions.

Forest Management: Indigenous peoples in the Amazon rainforest use knowledge of the forest ecosystem to practice sustainable harvesting and biodiversity conservation, contributing to forest resilience.

Integrating Indigenous Knowledge with Modern Science

Combining indigenous knowledge with modern scientific research can enhance climate resilience:

Collaborative Research: Partnerships between scientists and indigenous communities can lead to more effective and culturally appropriate climate adaptation strategies.

Community-Based Management: Empowering indigenous communities to manage their natural resources can lead to more sustainable and resilient outcomes.

Policy Integration: Recognizing and incorporating indigenous knowledge into national and international climate policies can ensure more holistic and inclusive approaches to climate resilience.

Challenges and Opportunities

Indigenous communities face numerous challenges, including land dispossession, cultural assimilation, and political marginalization. However, there are opportunities to support indigenous resilience:

Legal Recognition: Recognizing and protecting indigenous land rights and knowledge systems in national and international law.

Capacity Building: Providing resources and support for indigenous-led climate adaptation and mitigation projects.

Respect and Inclusion: Ensuring that indigenous voices and perspectives are included in climate decision-making processes.

Globalization and Cultural Homogenization

The Influence of Globalization

Globalization, the process by which economies, societies, and cultures become interconnected through trade, communication, and migration, has significant implications for cultural identity. While globalization can facilitate cultural exchange and economic development, it also poses risks of cultural homogenization, where local traditions and identities are overshadowed by dominant global cultures.

Cultural Homogenization and Loss of Diversity

Climate change can exacerbate the effects of globalization on cultural identity:

Displacement and Migration: Climate-induced displacement can lead to the mixing of cultures, which, while enriching, can also dilute distinct cultural identities.

Economic Pressures: Global economic systems can undermine traditional livelihoods and cultural practices, as communities are pressured to adopt more commercially viable activities.

Media and Communication: The global reach of media and communication technologies can spread dominant cultural narratives, marginalizing local languages, traditions, and identities.

Preserving Cultural Diversity

Efforts to preserve cultural diversity in the context of globalization and climate change include:

Cultural Revitalization: Supporting initiatives that revive and sustain local languages, arts, and traditions.

Economic Diversification: Encouraging economic activities that align with and support cultural preservation, such as eco-tourism and artisanal crafts.

Education and Awareness: Promoting education and awareness about the importance of cultural diversity and the impacts of climate change on cultural heritage.

Balancing Global and Local Identities

Finding a balance between embracing global interconnectedness and preserving local identities is crucial:

Hybrid Identities: Encouraging the development of hybrid identities that integrate global influences while maintaining a strong connection to local culture.

Cultural Exchange: Facilitating respectful and reciprocal cultural exchange that values and respects all cultural contributions.

Policy Support: Implementing policies that protect cultural diversity and support communities in maintaining their cultural heritage in the face of climate change and globalization.

This chapter explores the intricate relationship between climate change and cultural identity, highlighting the threats to cultural heritage, the value of indigenous knowledge, and the challenges of globalization. By understanding and addressing these impacts, we can better support the resilience and continuity of diverse cultural identities in a changing world.

Chapter 6
Climate Refugees Displacement and Identity Transformation

The Rise of Climate-Induced Migration

Understanding Climate Refugees

Climate refugees, also known as climate migrants, are individuals who are forced to leave their homes due to sudden or gradual environmental changes linked to climate change. These changes include rising sea levels, extreme weather events, prolonged droughts, and other environmental disruptions that make living conditions untenable. Unlike traditional refugees who flee due to conflict or persecution, climate refugees are driven by environmental factors.

Causes of Climate-Induced Migration

Rising Sea Levels: Coastal areas and island nations are particularly vulnerable to rising sea levels. Communities in places like the Maldives, Bangladesh, and Pacific Island nations are experiencing inundation of their land, leading to displacement.

Extreme Weather Events: Hurricanes, typhoons, and floods are becoming more frequent and severe, destroying homes and infrastructure. For example, hurricanes in the

Caribbean and cyclones in South Asia have displaced thousands.

Drought and Desertification: Prolonged droughts and desertification are rendering agricultural lands unproductive, forcing farmers and rural communities to migrate. The Sahel region in Africa is a prime example where desertification has driven migration.

Resource Scarcity: Changes in climate can lead to scarcity of essential resources like water and food, prompting migration in search of better living conditions.

Scale and Trends

The number of climate refugees is on the rise. According to the Internal Displacement Monitoring Centre (IDMC), millions of people are displaced annually due to climate-related disasters. This trend is expected to increase as the impacts of climate change intensify. The UN estimates that by 2050, there could be anywhere from 25 million to 1 billion climate refugees globally.

Case Studies

Bangladesh: One of the most vulnerable countries to climate change, Bangladesh experiences frequent flooding and cyclones. The rising sea levels are displacing coastal communities, forcing them to migrate inland.

Pacific Island Nations: Countries like Kiribati and Tuvalu face existential threats from rising sea levels. Entire communities are preparing to relocate, with some already seeking asylum in countries like New Zealand.

United States: In the US, extreme weather events such as hurricanes have led to internal displacement. For instance, Hurricane Katrina in 2005 displaced hundreds of thousands of people, many of whom struggled to return home or resettle permanently.

Identity in Transit: Redefining Home and Belonging

The Concept of Home

For many, home is more than just a physical place; it is a source of identity, belonging, and cultural connection. Displacement disrupts this sense of home, leading to profound psychological and social impacts. Climate refugees often face the challenge of redefining what home means as they move to new environments.

Psychological Impact of Displacement

Loss and Grief: The forced abandonment of home and community can lead to feelings of grief and loss. This emotional toll is compounded by the uncertainty and instability of displacement.

Identity Crisis: Displacement can result in an identity crisis as individuals struggle to reconcile their past lives with their new realities. This is particularly true for those whose cultural practices and livelihoods are tied to specific geographic locations.

Stress and Trauma: The journey of displacement and the conditions of resettlement can be stressful and traumatic. Climate refugees may face hazardous travel conditions, uncertain legal status, and hostile environments in their new locations.

Redefining Belonging

Building New Communities: Climate refugees often find themselves in new and unfamiliar places. Building new communities and social networks is crucial for regaining a sense of belonging.

Cultural Adaptation and Preservation: While adapting to new environments, climate refugees strive to preserve their cultural identities. This balance between adaptation and preservation is vital for maintaining a sense of self.

Intergenerational Challenges: The younger generation of climate refugees may face different identity challenges compared to their elders. While they might adapt more easily to new environments, they also risk losing touch with their cultural heritage.

Stories of Resilience

Kiribati Youth: Young people from Kiribati, facing the prospect of relocating due to rising sea levels, are engaging in cultural preservation efforts and advocacy, aiming to keep their traditions alive while adapting to new homes.

Bangladeshi Farmers: Displaced farmers from Bangladesh's coastal regions are finding ways to continue agricultural practices in new, less affected areas, ensuring food security and cultural continuity.

Policy and Human Rights: Protecting the Displaced

Legal and Policy Challenges

Current international frameworks do not adequately address the plight of climate refugees. Unlike traditional refugees, climate refugees lack specific legal protections under international law. The 1951 Refugee Convention does not recognize environmental factors as grounds for asylum.

The Need for Legal Recognition

Expanding Definitions: There is a growing call to expand the definition of refugees to include those displaced by climate change. This would provide legal recognition and protection for climate refugees.

National Policies: Countries need to develop national policies that address the specific needs of climate refugees. This includes creating pathways for relocation, integration, and providing necessary support services.

Human Rights Considerations

Climate refugees face numerous human rights challenges, including:

Right to Asylum: Ensuring that climate refugees have the right to seek asylum and are not returned to environments where their lives are at risk.

Right to Livelihood: Protecting the economic rights of climate refugees, including access to employment, education, and social services.

Right to Identity and Culture: Preserving the cultural identities and practices of displaced communities is crucial for maintaining their sense of self and community cohesion.

International Cooperation and Support

Addressing climate-induced displacement requires international cooperation:

Global Agreements: Strengthening global agreements, such as the Paris Agreement, to include provisions for supporting climate refugees.

Funding and Resources: Providing financial and technical support to countries and communities most affected by climate-induced displacement.

Capacity Building: Enhancing the capacity of local governments and organizations to manage and support displaced populations.

Case Studies of Policy and Support

New Zealand's Pacific Access Category: New Zealand has established a special visa category for citizens of Pacific Island nations affected by climate change, providing a legal pathway for relocation.

Germany's Climate Refugee Initiative: Germany has initiated programs to support climate refugees, including providing funding for adaptation and resilience projects in vulnerable regions.

This chapter delves into the multifaceted issue of climate refugees, exploring the rise of climate-induced migration, the psychological and social impacts of displacement, and the need for legal and policy frameworks to protect those affected.

Understanding and addressing these challenges is essential for ensuring the rights and dignity of climate refugees and supporting their transition to new lives and identities.

Chapter 7
Economic Inequality and Climate Change

The Rich and the Vulnerable: Disparate Impacts

Unequal Exposure and Vulnerability

Climate change does not affect all populations equally. Economic inequality exacerbates the impacts of climate change, with poorer communities and nations bearing the brunt of its effects. Wealthier individuals and countries have more resources to protect themselves and adapt to changing conditions, while vulnerable populations often lack the means to do so.

Geographic Disparities

Global South vs. Global North: Developing countries, particularly in the Global South, are more susceptible to climate change due to their geographical locations and limited resources. For instance, African nations face severe droughts, while small island developing states (SIDS) in the Pacific are threatened by rising sea levels.

Urban vs. Rural: Rural areas, which are often economically disadvantaged, face greater risks from climate impacts like agricultural disruption and water scarcity. Urban areas, while also vulnerable, often have better infrastructure and resources for adaptation.

Socioeconomic Factors

Income Levels: Low-income households are less likely to have access to resources that can mitigate the effects of climate change, such as air conditioning during heatwaves or insurance to cover damages from natural disasters.

Housing and Infrastructure: Poorer communities often live in areas with substandard housing and inadequate infrastructure, making them more susceptible to flooding, landslides, and other climate-related events.

Health Disparities: Economic inequality affects health outcomes, with poorer populations suffering more from climate-related health issues such as heat stress, malnutrition, and vector-borne diseases.

Case Studies

Hurricane Katrina: In 2005, Hurricane Katrina devastated New Orleans, disproportionately affecting low-income and African American communities. These populations faced greater difficulties in evacuating, had less access to resources for recovery, and experienced prolonged displacement.

Bangladesh: Frequent flooding and cyclones in Bangladesh disproportionately impact the country's poor, who often live in low-lying, flood-prone areas. The economic losses from these events further entrench poverty.

Climate Justice: Bridging Economic Divides

Principles of Climate Justice

Climate justice addresses the ethical and political dimensions of climate change, emphasizing the need to protect the rights and wellbeing of the most vulnerable populations. It calls for fair and equitable distribution of the burdens and benefits of climate action.

Historical Responsibility

Carbon Emissions: Developed nations have historically contributed the most to greenhouse gas emissions, yet developing countries are the ones experiencing the most severe impacts. Climate justice argues for acknowledging this historical responsibility and providing support to those disproportionately affected.

Equity in Climate Action

Adaptation and Mitigation Funding: Ensuring that financial and technical support for climate adaptation and mitigation is directed towards the most vulnerable communities. This includes international funding mechanisms such as the Green Climate Fund.

Inclusive Policy Making: Involving marginalized and vulnerable communities in the decision-making process to ensure that their voices are heard and their needs addressed.

Strategies for Climate Justice

Loss and Damage Mechanisms: Establishing frameworks to compensate communities and countries for the irreversible impacts of climate change, such as loss of land and livelihoods.

Technology Transfer: Facilitating the transfer of clean and resilient technologies to developing countries to help them adapt to and mitigate climate change.

Capacity Building: Enhancing the capabilities of vulnerable communities to manage climate risks through education, training, and infrastructure development.

Case Studies

UNFCCC's Warsaw International Mechanism: This mechanism addresses loss and damage associated with the impacts of climate change in developing countries that are particularly vulnerable.

Green Climate Fund: A global initiative aimed at supporting developing countries in their efforts to combat climate change through funding projects that promote low-emission and climate-resilient development.

Redefining Success: Sustainable Economies

Moving Beyond GDP

Traditional economic measures like Gross Domestic Product (GDP) often fail to account for environmental degradation and social wellbeing. Redefining economic success involves integrating sustainability and human welfare into economic indicators and policies.

Alternative Metrics

Gross National Happiness (GNH): Adopted by Bhutan, GNH measures the collective happiness and wellbeing of a population, incorporating environmental and social factors.

Human Development Index (HDI): Developed by the United Nations, HDI combines indicators of health, education, and income to provide a broader measure of development.

Genuine Progress Indicator (GPI): This metric adjusts GDP by including environmental and social costs, offering a more comprehensive view of economic progress.

Sustainable Economic Models

Circular Economy: This model emphasizes reusing, recycling, and regenerating materials to create a closed-loop system, reducing waste and environmental impact. It promotes sustainable production and consumption.

Green Economy: Focuses on reducing carbon emissions, increasing energy efficiency, and promoting sustainable practices in all sectors of the economy. It aims to achieve sustainable development without degrading the environment.

Social Enterprise: Businesses that prioritize social and environmental goals alongside profit, contributing to community development and sustainability.

Policy and Practice

Regulatory Frameworks: Governments can implement policies that promote sustainability, such as carbon pricing, renewable energy incentives, and environmental regulations.

Corporate Responsibility: Businesses can adopt sustainable practices, such as reducing their carbon footprint, using sustainable materials, and engaging in corporate social responsibility (CSR) initiatives.

Community-Led Initiatives: Empowering communities to develop and implement their own sustainable practices can lead to more resilient and equitable outcomes.

Case Studies

Costa Rica: Known for its commitment to sustainability, Costa Rica has achieved significant progress in renewable energy, reforestation, and conservation, demonstrating the potential of a green economy.

Germany's Energiewende: Germany's energy transition strategy aims to shift from fossil fuels to renewable energy sources, reducing carbon emissions and promoting energy efficiency.

This chapter explores the intersection of economic inequality and climate change, highlighting the disproportionate impacts on vulnerable populations and the need for climate justice. It also emphasizes the importance of redefining economic success through sustainable and inclusive economic models. By addressing economic disparities and promoting equitable climate action, we can create a more just and sustainable future for all.

Chapter 8
Gender and Climate Change

Gendered Impacts: Women and Climate Vulnerability

Disproportionate Vulnerability

Climate change affects women and men differently due to existing social, economic, and cultural inequalities. Women, particularly in developing countries, often bear the brunt of climate impacts because they have less access to resources, face restrictive social norms, and are often responsible for securing food, water, and energy for their families.

Economic Inequality

Employment and Income: Women are more likely to be employed in informal and low-paying jobs, which lack social protections and are highly vulnerable to climate disruptions. For example, women working in agriculture or as small-scale traders face significant economic losses during climate-related disasters.

Asset Ownership: Women typically own fewer assets such as land, livestock, and property. This lack of asset ownership limits their ability to recover from climate-induced losses and increases their economic vulnerability.

Health and Safety

Health Risks: Women are more susceptible to health issues arising from climate change, such as malnutrition, waterborne diseases, and heat stress. For instance, pregnant women face heightened health risks during extreme weather events.

Safety and Security: During and after climate disasters, women are at greater risk of gender-based violence, including domestic violence, sexual assault, and human trafficking. Displacement and overcrowded shelters often exacerbate these risks.

Social and Cultural Norms

Caregiving Roles: Women are often the primary caregivers in families, responsible for children, the elderly, and the sick. Climate change can increase caregiving burdens, as women must travel further for water and firewood, or take on additional care responsibilities due to health impacts.

Mobility and Decision-Making: Social norms can restrict women's mobility and limit their participation in decision-making processes related to climate adaptation and disaster response. This marginalization reduces their ability to access relief and recovery resources.

Case Studies

Bangladesh: In Bangladesh, women are disproportionately affected by flooding and cyclones. Traditional norms often restrict their movement, making it harder for them to evacuate in emergencies. Women also face increased health risks and economic hardships due to the destruction of crops and livestock.

Sub-Saharan Africa: In many parts of sub-Saharan Africa, women are responsible for collecting water and firewood. Droughts and desertification force them to travel longer distances, increasing their exposure to physical danger and reducing time for other productive activities.

Empowering Women: Agents of Change

Women as Key Stakeholders

Despite their vulnerability, women play a crucial role in addressing climate change. Their unique knowledge, skills, and perspectives make them essential agents of change in building climate resilience and promoting sustainable development.

Leadership and Participation

Community Leadership: Women's involvement in community leadership and decision-making processes enhances the effectiveness of climate adaptation and disaster risk reduction strategies. Women-led initiatives often prioritize the needs of the most vulnerable and ensure more inclusive and equitable solutions.

Political Participation: Increasing women's representation in political institutions at local, national, and international levels is critical for integrating gender-sensitive approaches into climate policies and programs.

Education and Capacity Building

Education and Awareness: Educating women and girls about climate change and environmental sustainability empowers them to take action and advocate for their rights. Education also improves their ability to access information and resources for adaptation.

Capacity Building: Providing training and resources to women in areas such as sustainable agriculture, renewable energy, and disaster preparedness enhances their capacity to respond to climate challenges and contribute to community resilience.

Sustainable Livelihoods

Agriculture and Food Security: Supporting women farmers with access to climate-resilient crops, sustainable farming techniques, and financial services can improve food security and economic stability. Women's traditional knowledge in seed saving and soil management is invaluable for adapting to changing conditions.

Renewable Energy: Promoting women's involvement in the renewable energy sector, including as entrepreneurs and technicians, creates economic opportunities and reduces dependence on fossil fuels. Women-led initiatives in solar energy, for example, provide affordable and sustainable energy solutions to rural communities.

Case Studies

India's Self-Help Groups: In India, women's self-help groups have been instrumental in promoting sustainable agriculture, water management, and disaster preparedness. These groups provide a platform for women to share knowledge, access microfinance, and advocate for community needs.

Kenya's Green Belt Movement: Founded by Wangari Maathai, the Green Belt Movement has mobilized women to plant millions of trees, restoring degraded environments, improving livelihoods, and empowering women through environmental conservation.

Intersectionality: Understanding Diverse Experiences

The Concept of Intersectionality

Intersectionality is a framework for understanding how various aspects of a person's social and political identities (e.g., gender, race, class, ethnicity, sexuality, disability) intersect and contribute to unique experiences of oppression and privilege. In the context of climate change, intersectionality highlights how overlapping identities shape individuals' vulnerability and capacity to respond to climate impacts.

Multiple Layers of Disadvantage

Race and Ethnicity: Women from racial and ethnic minority groups often face compounded disadvantages due to systemic racism and discrimination. These factors can limit their access to resources, reduce their socioeconomic opportunities, and increase their exposure to climate risks.

Economic Status: Poor women are more vulnerable to climate change due to limited financial resources, reduced access to education and healthcare, and insecure housing. Economic status intersects with gender to create heightened levels of vulnerability.

Disability: Women with disabilities face additional barriers in accessing climate information, participating in decision-making, and evacuating during disasters. They are often overlooked in climate policies and programs, exacerbating their marginalization.

Inclusive Approaches

Targeted Interventions: Developing targeted interventions that address the specific needs of women with intersecting identities is essential for effective climate action. This includes ensuring accessibility of disaster relief services, providing tailored support for minority and indigenous women, and addressing the unique challenges faced by LGBTQ+ communities.

Participatory Processes: Involving women from diverse backgrounds in participatory processes ensures that climate policies and programs are inclusive and equitable. Engaging marginalized women in planning and implementation can lead to more effective and just outcomes.

Case Studies

Indigenous Women in the Amazon: Indigenous women in the Amazon play a crucial role in protecting forests and biodiversity. Their traditional knowledge and leadership are essential for sustainable resource management. However, they face multiple layers of marginalization, including gender discrimination and threats to their land rights.

Urban Poor Women in the Philippines: In urban areas of the Philippines, poor women are disproportionately affected by climate impacts such as flooding and typhoons. Intersectional approaches that address their housing, health, and livelihood needs are vital for building urban resilience.

Policy Recommendations

Inclusive Policy Design: Policymakers should adopt an intersectional lens when designing climate policies to ensure they address the diverse needs of all women. This includes collecting disaggregated data, conducting gender and social impact assessments, and ensuring representation of marginalized groups.

Integrated Social Protection: Integrating social protection measures such as cash transfers, healthcare access, and housing support with climate adaptation strategies can enhance the resilience of women with intersecting vulnerabilities.

Community-Based Approaches: Supporting community-based approaches that empower women and incorporate their knowledge and leadership can lead to more sustainable and equitable climate solutions.

This chapter explores the complex relationship between gender and climate change, highlighting the disproportionate impacts on women, the critical role of women as agents of change, and the importance of intersectional approaches to understanding diverse experiences. By addressing gender inequalities and empowering women, we can build more resilient and inclusive societies capable of effectively responding to climate challenges.

Chapter 9
Youth Identity in the Age of Climate Change

The Climate Generation: Youth Activism

Rise of the Climate Generation

The current generation of young people is increasingly identified as the "Climate Generation" due to their heightened awareness of and activism against climate change. This cohort is uniquely positioned in a world where the impacts of climate change are more visible and urgent than ever before. Their identity is being shaped by their proactive stance in the fight against climate change, driven by a deep sense of urgency and responsibility.

Global Movements

Fridays for Future: Initiated by Greta Thunberg, this movement has inspired millions of young people worldwide to strike from school on Fridays to demand action on climate change. It highlights the collective power of youth to influence public opinion and policy.

Sunrise Movement: In the United States, the Sunrise Movement, led by young people, advocates for political action on climate change, including the Green New Deal. Their activism emphasizes the intersection of climate action with social justice.

Motivations and Concerns

Existential Threat: Young people view climate change as an existential threat that will define their future. This perspective drives their activism, as they fight for a livable planet.

Intergenerational Justice: Youth activists often frame their demands within the context of intergenerational justice, arguing that older generations have a moral obligation to address the climate crisis they have contributed to.

Impact of Youth Activism

Policy Influence: Youth movements have succeeded in bringing climate change to the forefront of political agendas. Their persistent activism has pressured governments to commit to more ambitious climate policies.

Social Awareness: The activism of young people has significantly raised awareness about climate change, particularly among their peers. They use social media and digital platforms to disseminate information, mobilize support, and engage in global dialogues.

Case Studies

Greta Thunberg: Greta's solo protest outside the Swedish parliament sparked a global movement, earning her recognition as a leading voice in the climate movement. Her speeches at international forums have galvanized global youth activism.

Autumn Peltier: An indigenous water activist from Canada, Autumn advocates for clean water and environmental protection. She represents the intersection of youth activism and indigenous rights, highlighting the role of young indigenous leaders in the climate movement.

Future Prospects: Education and Career Shifts

Transforming Education

As climate change becomes a defining issue of our time, education systems are evolving to prepare young people for a future where environmental challenges are central. This shift is reshaping the educational landscape and influencing career aspirations.

Integrating Climate Education

Curriculum Changes: Schools and universities are incorporating climate science, sustainability, and environmental studies into their curricula. This includes interdisciplinary approaches that connect climate issues with subjects like economics, politics, and ethics.

Experiential Learning: Programs that involve hands-on experiences, such as community projects, internships, and fieldwork, are becoming more common. These opportunities allow students to engage directly with environmental issues and solutions.

Career Shifts

Green Jobs: The demand for green jobs is increasing as industries and governments prioritize sustainability. Careers in renewable energy, environmental engineering, conservation, and sustainable agriculture are becoming more prominent.

Social Entrepreneurship: Many young people are drawn to social entrepreneurship, starting businesses that address environmental and social challenges. These ventures often focus on innovative solutions to reduce carbon footprints, promote circular economies, and enhance community resilience.

Skill Development

Technical Skills: Proficiency in technical skills related to renewable energy, green technology, and environmental management is becoming essential. Educational institutions are responding by offering specialized training and certifications.

Soft Skills: Critical thinking, problem-solving, collaboration, and leadership skills are increasingly valued. These skills enable young people to navigate complex environmental issues and lead effective climate action.

Case Studies

Green School Bali: This innovative school in Indonesia integrates sustainability into every aspect of its curriculum, from academic subjects to practical skills like organic farming and waste management.

Eco-Incubators: Programs like the Climate-KIC initiative in Europe support young entrepreneurs in developing sustainable business ideas, providing mentorship, funding, and networking opportunities.

Building a Sustainable Identity: Youth Leadership

Emerging Leaders

Young people are not only advocating for climate action but also stepping into leadership roles that drive tangible change. Their leadership is characterized by creativity, inclusivity, and a strong commitment to sustainability.

Youth-led Organizations

Global Youth Biodiversity Network (GYBN): This international network of youth organizations works to promote biodiversity conservation. It empowers young leaders to engage in policy discussions and implement local conservation projects.

Plant-for-the-Planet: Founded by Felix Finkbeiner at the age of nine, this organization mobilizes young people worldwide to plant trees and combat climate change. It demonstrates the power of youth-led environmental initiatives.

Innovative Solutions

Tech and Innovation: Young innovators are developing technological solutions to address climate challenges. From apps that track carbon footprints to drones that monitor forest health, youth-led tech initiatives are making significant contributions.

Community Initiatives: Many young leaders are focused on grassroots efforts that build local resilience. These initiatives often involve community education, sustainable farming practices, and the creation of green spaces.

Mentorship and Networks

Intergenerational Collaboration: Young leaders benefit from mentorship and collaboration with experienced professionals and activists. These relationships provide guidance, resources, and opportunities for scaling impact.

Youth Networks: Networks like Youth Climate Leaders (YCL) connect young climate activists globally, offering training, exchange programs, and support for project implementation. These networks foster a sense of community and collective action.

Case Studies

Leah Namugerwa: A Ugandan climate activist, Leah leads campaigns for tree planting and plastic bag bans in her country. Her leadership exemplifies the impact of youth activism on local policy and environmental practices.

Boyan Slat: As the founder of The Ocean Cleanup, Boyan has led efforts to develop technologies to remove plastic pollution from the world's oceans. His work highlights the role of young innovators in addressing global environmental issues.

This chapter explores the profound impact of climate change on youth identity, highlighting the rise of youth activism, the transformation of education and career prospects, and the emergence of young leaders driving sustainable change. By embracing their role as agents of change and leveraging their unique skills and perspectives, young people are building a sustainable identity that will shape the future of our planet.

Chapter 10
Technological Identity in a Warming World

Innovation and Adaptation: Technological Solutions

Cutting-Edge Innovations

Technology plays a pivotal role in addressing climate change by offering innovative solutions for mitigation and adaptation. As the climate crisis intensifies, the development and deployment of advanced technologies are crucial for reducing greenhouse gas emissions and enhancing resilience to climate impacts.

Renewable Energy

Solar and Wind Power: Advances in solar and wind technologies have significantly reduced the cost of renewable energy, making it more accessible and scalable. Innovations such as bifacial solar panels and floating wind turbines are further enhancing efficiency and expanding deployment options.

Energy Storage: Breakthroughs in battery technology, such as lithium-ion and emerging solid-state batteries, are improving energy storage capabilities. This allows for better integration of intermittent renewable energy sources into the grid, ensuring a stable and reliable energy supply.

Smart Grids and Energy Efficiency

Smart Grids: The development of smart grids, which use digital technology to monitor and manage electricity flows, enhances grid reliability and efficiency. Smart grids enable better integration of renewable energy sources and facilitate demand response programs.

Energy Efficiency: Innovations in building materials, lighting, and heating and cooling systems are reducing energy consumption. Technologies such as smart thermostats, energy-efficient appliances, and advanced insulation are contributing to more sustainable living environments.

Climate Adaptation Technologies

Climate-Resilient Agriculture: Technologies like precision farming, drought-resistant crops, and automated irrigation systems help farmers adapt to changing climate conditions. These innovations improve crop yields and reduce water usage, ensuring food security in the face of climate change.

Flood and Disaster Management: Early warning systems, flood barriers, and resilient infrastructure are critical for protecting communities from extreme weather events. Technologies such as drones, remote sensing, and geographic information systems (GIS) enhance disaster preparedness and response.

Carbon Capture and Storage (CCS)

Carbon Sequestration: Technologies for capturing and storing carbon dioxide (CO_2) from industrial processes and power plants are essential for mitigating climate change.

Techniques such as direct air capture and bioenergy with carbon capture and storage (BECCS) are being developed to reduce atmospheric CO_2 levels.

Enhanced Oil Recovery: Some CCS technologies are used in enhanced oil recovery, where captured CO_2 is injected into oil fields to increase extraction efficiency. While this has raised ethical concerns, it also provides a potential revenue stream for CCS projects.

Case Studies

Tesla Powerwall: Tesla's Powerwall battery system stores solar energy for residential use, providing backup power and enabling homes to become energy self-sufficient.

Vertical Farming: Companies like AeroFarms use vertical farming technology to grow crops in controlled indoor environments, reducing water usage and eliminating the need for pesticides.

Digital Activism: Online Movements and Climate Awareness

The Power of Digital Platforms

Digital technology has revolutionized activism by providing new tools for organizing, mobilizing, and raising awareness about climate change. Online platforms enable activists to reach a global audience, share information rapidly, and coordinate actions more effectively.

Social Media Campaigns

Hashtag Movements: Hashtags like #ClimateStrike, #FridaysForFuture, and #ActOnClimate have become rallying points for online climate activism. These digital campaigns amplify voices, spread awareness, and encourage participation in climate actions.

Viral Content: Videos, infographics, and memes are powerful tools for communicating complex climate issues in an engaging and accessible manner. Viral content can raise awareness, inspire action, and put pressure on policymakers and corporations.

Online Communities and Collaboration

Virtual Communities: Online forums, social media groups, and digital platforms connect climate activists from around the world. These virtual communities provide support, share resources, and foster collaboration on climate initiatives.

Crowdsourcing Solutions: Digital platforms enable crowdsourcing of ideas and solutions for climate challenges. Initiatives like hackathons and online competitions encourage innovation and collective problem-solving.

E-Petitions and Digital Advocacy

Online Petitions: Platforms like Change.org and Avaaz allow activists to create and sign online petitions, demanding action from governments and corporations. These petitions can gather millions of signatures and exert significant influence.

Digital Advocacy: Email campaigns, social media advocacy, and online lobbying tools empower individuals to contact their representatives, advocate for policy changes, and participate in public consultations.

Case Studies

Extinction Rebellion: This global movement uses digital platforms to coordinate nonviolent direct actions and civil disobedience campaigns, raising awareness about the climate emergency.

Greta Thunberg's Social Media: Greta Thunberg's use of social media to document her climate strikes and speeches has inspired millions of young people to join the climate movement.

Ethical Considerations: Technology and Environmental Impact

Balancing Innovation with Sustainability

While technology offers significant benefits in addressing climate change, it also raises ethical concerns related to its environmental impact, social implications, and equitable access. Balancing technological innovation with sustainability and ethical considerations is crucial.

Environmental Footprint

Resource Extraction: The production of technologies like solar panels, wind turbines, and batteries requires the extraction of raw materials such as rare earth metals and lithium. Mining practices can lead to environmental degradation, habitat destruction, and pollution.

E-Waste: The rapid pace of technological advancement leads to increased electronic waste (e-waste). Proper disposal and recycling of e-waste are essential to prevent environmental contamination and resource depletion.

Social and Economic Impacts

Equitable Access: Ensuring equitable access to climate technologies is critical for addressing global disparities. Developing countries and marginalized communities often face barriers in accessing and benefiting from advanced technologies.

Job Displacement: The transition to a low-carbon economy can lead to job displacement in industries like fossil fuels. Policies and programs that support retraining and job creation in sustainable sectors are necessary to mitigate social impacts.

Ethical Use of Data

Data Privacy: The collection and use of data for climate monitoring, smart grids, and digital activism raise concerns about data privacy and security. Ensuring that data is used ethically and that individuals' privacy is protected is essential.

Surveillance: Technologies used for climate monitoring, such as drones and remote sensing, can also be used for surveillance. Safeguarding against misuse and ensuring transparency and accountability are important ethical considerations.

Responsible Innovation

Inclusive Innovation: Involving diverse stakeholders, including marginalized communities, in the development and deployment of climate technologies ensures that innovations address their needs and are culturally appropriate.

Sustainable Design: Designing technologies with sustainability in mind, from the materials used to the end-of-life disposal, minimizes their environmental impact. Circular economy principles, such as designing for reuse and recycling, are key to sustainable innovation.

Case Studies

Fairphone: Fairphone produces modular smartphones designed for longevity, repairability, and ethical sourcing of materials, addressing both environmental and social impacts of technology.

Lithium Extraction in Bolivia: Bolivia's approach to lithium extraction, emphasizing environmental protection and fair labor practices, illustrates the importance of ethical resource management in the production of climate technologies.

This chapter explores the intersection of technology and climate change, highlighting the innovative solutions driving mitigation and adaptation efforts, the power of digital activism in raising climate awareness, and the ethical considerations that must be addressed to ensure sustainable and equitable technological development.

By embracing responsible innovation and leveraging digital platforms, we can harness the power of technology to build a more resilient and sustainable future.

Chapter 11
Climate Change and National Identity

National Narratives: Climate Change in Policy and Media

Shaping National Discourse

The way a nation addresses climate change is deeply intertwined with its national identity. Policies, media coverage, and public narratives reflect and shape how a country perceives itself in the context of the global climate crisis.

Climate Policy as National Identity

Policy Frameworks: Nations often craft climate policies that align with their broader identity. For instance, Scandinavian countries like Sweden and Denmark pride themselves on progressive environmental policies, viewing themselves as global leaders in sustainability.

Legislation and Targets: Ambitious climate targets and legislation, such as net-zero emissions goals, are increasingly seen as markers of national identity. Countries that adopt aggressive climate policies can bolster their international reputation and influence.

Media Representation

Framing the Issue: The media plays a crucial role in shaping public perception of climate change. How climate issues are framed—whether as a national security threat, an economic opportunity, or a moral imperative—affects national identity and policy responses.

Public Discourse: Media coverage influences public discourse and, consequently, the political climate. In countries where media emphasizes the urgency of climate action, public support for strong policies tends to be higher.

Cultural Narratives

Historical Context: National narratives around climate change often draw on historical experiences. For example, nations with a history of environmentalism may emphasize their legacy in conservation and stewardship.

Cultural Values: Countries may integrate climate change into their cultural values, such as Japan's emphasis on harmony with nature or New Zealand's Māori principles of kaitiakitanga (guardianship).

Case Studies

Germany's Energiewende: Germany's energy transition policy (Energiewende) reflects its commitment to renewable energy and sustainability. This policy is central to its national identity as a leader in green technology.

China's Climate Strategy: China frames its climate policies within the context of modernization and economic development. While balancing environmental goals with growth, China aims to position itself as a responsible global power.

Sovereignty and Responsibility: International Dynamics

Balancing National Interests and Global Obligations

Climate change challenges the traditional notions of sovereignty, as its impacts and solutions transcend national borders. Countries must navigate the tension between protecting national interests and fulfilling international responsibilities.

International Agreements

Paris Agreement: The Paris Agreement is a landmark accord that brings nations together to combat climate change. While it respects national sovereignty by allowing countries to set their own targets, it also requires them to contribute to a global effort.

Kyoto Protocol: The Kyoto Protocol established legally binding targets for developed countries to reduce greenhouse gas emissions. It highlighted the principle of "common but differentiated responsibilities," recognizing different capabilities and responsibilities among nations.

Sovereignty Concerns

Resource Management: Nations often face dilemmas over how to manage natural resources in the context of climate change. Sovereignty concerns can lead to conflicts over water rights, land use, and energy production.

Economic Interests: Balancing economic growth with climate commitments is a significant challenge. Developing nations, in particular, may prioritize economic development over stringent climate measures, arguing for their right to industrialize.

Global Solidarity

Climate Justice: The concept of climate justice emphasizes that those who have contributed least to climate change are often most affected. Developed countries are called to support developing nations through finance, technology transfer, and capacity-building.

International Cooperation: Effective climate action requires unprecedented levels of international cooperation. Multilateral forums like the United Nations Framework Convention on Climate Change (UNFCCC) facilitate dialogue and collaboration among nations.

Case Studies

Small Island Developing States (SIDS): SIDS are among the most vulnerable to climate change impacts like sea-level rise and extreme weather. They advocate strongly for international support and action, highlighting the disparity between their low emissions and high vulnerability.

European Union (EU): The EU has positioned itself as a global leader in climate policy, with stringent emissions targets and a comprehensive climate strategy. The EU's approach emphasizes regional cooperation and collective action.

National Pride vs. Global Responsibility: Balancing Acts

Reconciling National Pride with Global Commitment

Nations face the complex task of balancing national pride and sovereignty with the need for global cooperation and responsibility in addressing climate change.

National Interests

Economic Competitiveness: Countries often frame their climate policies in terms of economic competitiveness, promoting green industries as a source of national pride and job creation.

Energy Independence: Reducing reliance on imported fossil fuels through renewable energy is presented as a matter of national security and pride, enhancing energy independence.

Global Responsibilities

Emission Reductions: As major emitters, countries like the United States, China, and India have significant responsibilities to reduce their greenhouse gas emissions. Their actions are crucial for global climate targets.

Climate Finance: Wealthier nations are expected to provide financial support to developing countries to help them mitigate and adapt to climate change. This includes contributions to the Green Climate Fund and other mechanisms.

Public Perception and Policy

Nationalism and Climate Action: Nationalist sentiments can both support and hinder climate action. While some leaders leverage national pride to promote environmental stewardship, others may resist international agreements perceived as limiting national sovereignty.

Citizen Engagement: Public support is essential for ambitious climate policies. Governments that effectively engage their citizens, emphasizing both national benefits and global responsibilities, are more likely to succeed in implementing robust climate measures.

Case Studies

United States: The U.S. has experienced fluctuating climate policies influenced by changes in administration. Balancing economic interests, energy independence, and international commitments remains a contentious issue.

Norway: Norway combines national pride in its oil industry with a commitment to environmental responsibility. It invests heavily in renewable energy and climate finance, striving to reconcile its economic interests with global climate goals.

Policy Recommendations

Integrative Approaches: Policymakers should adopt integrative approaches that align national interests with global responsibilities. This includes crafting policies that simultaneously promote economic growth, national security, and environmental sustainability.

Public Awareness Campaigns: Governments should invest in public awareness campaigns that emphasize the interconnectedness of national well-being and global climate action. Educating citizens about the benefits of climate policies can build broad support.

International Leadership: Nations should strive to lead by example, demonstrating how strong climate action can enhance national pride and global standing. Effective leadership can inspire other countries to follow suit, fostering a collaborative international climate effort.

This chapter delves into the intricate relationship between climate change and national identity, exploring how countries shape their narratives, navigate sovereignty and responsibility, and balance national pride with global commitments. By understanding and addressing these dynamics, nations can contribute to a more effective and unified global response to the climate crisis.

Chapter 12
Religion, Spirituality, and Climate Change

Faith-Based Responses: Religious Environmentalism

Religious Teachings and Environmental Stewardship

Many world religions have teachings that emphasize the importance of caring for the Earth. These teachings form the basis of faith-based responses to climate change, encouraging adherents to act as stewards of the environment.

Christianity

Stewardship Doctrine: The concept of stewardship in Christianity teaches that humans are caretakers of God's creation. This doctrine motivates many Christian groups to engage in environmental conservation and climate action.

Pope Francis's Laudato Si': In his encyclical *Laudato Si'*, Pope Francis calls for urgent action to combat climate change and protect the environment, framing it as a moral and spiritual imperative.

Islam

Khalifa (Stewardship): Islam emphasizes the role of humans as stewards (khalifa) of the Earth. The Qur'an and Hadith provide numerous references to the importance of protecting the environment.

Green Ramadan Initiatives: Some Muslim communities have initiated Green Ramadan projects, promoting eco-friendly practices during the holy month, such as reducing waste and conserving water.

Buddhism

Interconnectedness: Buddhist teachings emphasize the interconnectedness of all life forms and the importance of compassion. This perspective fosters a deep respect for nature and a commitment to environmental preservation.

Engaged Buddhism: Movements like Engaged Buddhism, led by figures like Thich Nhat Hanh, integrate social and environmental activism with spiritual practice, advocating for sustainable living.

Hinduism

Dharmic Responsibility: Hinduism teaches that it is dharma (duty) to protect nature. The principles of ahimsa (non-violence) and karma (actions and their consequences) support environmental sustainability.

Sacred Rivers and Forests: The reverence for natural elements like rivers (e.g., Ganges) and forests in Hinduism encourages conservation efforts to protect these sacred sites from pollution and degradation.

Faith-Based Environmental Movements

Interfaith Power and Light (IPL): This organization mobilizes religious communities in the United States to take action on climate change through energy conservation, renewable energy adoption, and advocacy.

The Green Church Movement: Many Christian denominations participate in the Green Church Movement, implementing sustainable practices within their congregations and advocating for environmental policies.

Case Studies

The Ecumenical Patriarch Bartholomew I: Known as the "Green Patriarch," Bartholomew I of the Eastern Orthodox Church has been a vocal advocate for environmental protection, linking ecological issues with spiritual responsibility.

Islamic Foundation for Ecology and Environmental Sciences (IFEES): IFEES works to promote environmental awareness and action within Muslim communities, offering educational resources and practical guidance on sustainable practices.

Spiritual Identity: Connecting with Nature

Nature as a Spiritual Experience

For many individuals, nature is a profound source of spiritual connection and identity. This relationship fosters a deep sense of responsibility for the environment and motivates climate action.

Nature and Spiritual Practices

Meditation and Reflection: Many spiritual traditions incorporate practices that connect individuals with nature, such as meditation in natural settings, forest bathing (shinrin-yoku), and pilgrimage to sacred natural sites.

Rituals and Ceremonies: Rituals that honor the natural world, such as solstice celebrations, earth-based rituals in neo-pagan traditions, and Native American ceremonies, reinforce the spiritual significance of nature.

Environmental Ethics

Deep Ecology: This philosophical movement promotes the intrinsic value of all living beings and the need for a radical shift in human consciousness to foster harmony with nature.

Eco-Spirituality:

Eco-spirituality combines environmentalism with spiritual beliefs, emphasizing the sacredness of the Earth and the need for sustainable living. It is a growing movement among various religious and spiritual communities.

Personal Transformation

Healing and Well-being: Connection with nature is often associated with physical, mental, and spiritual well-being. Activities like hiking, gardening, and spending time in green spaces can enhance personal health and foster a deeper appreciation for the environment.

Spiritual Awakening: Experiences in nature can lead to spiritual awakening and a profound sense of interconnectedness, prompting individuals to adopt more sustainable lifestyles and advocate for environmental protection.

Case Studies

The Pachamama Alliance: This organization, inspired by the indigenous wisdom of the Achuar people of the Amazon, promotes a holistic approach to environmental activism, integrating spiritual awareness with ecological sustainability.

Joanna Macy's Work That Reconnects: Environmental activist and Buddhist scholar Joanna Macy's framework, "The Work That Reconnects," facilitates personal and collective transformation through experiential practices that deepen the connection with nature and inspire action.

Interfaith Dialogue: Unified Climate Action

Collaboration Across Faiths

Interfaith dialogue on climate change brings together diverse religious communities to address a common existential threat. By finding common ground and working collaboratively, these efforts amplify the impact of climate action.

Building Common Ground

Shared Values: Many religions share core values such as stewardship, compassion, and justice, which can serve as a foundation for collaborative climate action. Recognizing these shared values fosters unity and cooperation.

Joint Statements and Declarations: Interfaith declarations, such as the Interfaith Climate Change Statement to World Leaders, call for urgent action and express a unified moral and spiritual stance on climate issues.

Interfaith Organizations and Initiatives

The Interfaith Rainforest Initiative: This global alliance of religious leaders and faith-based organizations works to protect rainforests and the rights of indigenous peoples, recognizing the spiritual and ecological importance of these ecosystems.

GreenFaith: GreenFaith is an international interfaith coalition that inspires and mobilizes people of diverse religious backgrounds to take environmental action through advocacy, education, and sustainable practices.

Challenges and Opportunities

Cultural Differences: While interfaith collaboration offers significant potential, it also faces challenges such as cultural differences and varying theological perspectives. Effective dialogue requires mutual respect and a willingness to learn from one another.

Strengthening Solidarity: Successful interfaith initiatives build solidarity and foster a sense of global community. By working together, religious groups can leverage their collective influence to advocate for stronger climate policies and promote sustainable practices.

Policy and Advocacy

Moral Leadership: Religious leaders have a unique ability to influence public opinion and policy. By framing climate action as a moral and spiritual imperative, they can mobilize their followers and advocate for meaningful change at local, national, and international levels.

Grassroots Mobilization: Faith-based organizations often have extensive grassroots networks, which can be mobilized to support environmental campaigns, participate in climate marches, and implement sustainable practices within communities.

Case Studies

The Parliament of the World's Religions: This global interfaith organization has made climate change a central focus, promoting dialogue and collaboration among diverse religious traditions to address environmental challenges.

The Climate Action Network (CAN): CAN's interfaith caucus brings together representatives from various religious communities to advocate for ambitious climate policies at international climate negotiations.

This chapter explores the multifaceted relationship between religion, spirituality, and climate change, highlighting faith-based responses, the spiritual connection with nature, and the power of interfaith dialogue. By embracing their spiritual values and working collaboratively, religious and spiritual communities can play a vital role in addressing the climate crisis and fostering a more sustainable and just world.

Chapter 13
Art and Expression in the Climate Crisis

Climate Change in Literature and Film

Literature: Imagining the Climate Future

Climate change has become a powerful theme in contemporary literature, with authors exploring its impact on society, the environment, and human psychology. This genre, often referred to as "climate fiction" or "cli-fi," helps readers envision possible futures and grapple with the ethical and existential questions posed by the climate crisis.

Climate Fiction (Cli-Fi)

Dystopian Futures: Many cli-fi novels present dystopian futures where climate change has led to societal collapse, resource scarcity, and environmental degradation. These narratives often serve as cautionary tales, warning readers about the consequences of inaction.

Human Resilience: Other cli-fi works focus on human resilience and adaptation, highlighting the ways in which individuals and communities respond to climate challenges. These stories can inspire hope and motivate action.

Notable Works

"The Road" by Cormac McCarthy: This Pulitzer Prize-winning novel depicts a post-apocalyptic world ravaged by an unspecified catastrophe, often interpreted as climate-related. It explores themes of survival, morality, and the enduring bond between father and son.

"Parable of the Sower" by Octavia Butler: Set in a future America devastated by climate change and societal collapse, Butler's novel follows a young woman who develops a new belief system and leads a community towards hope and resilience.

"Flight Behavior" by Barbara Kingsolver: This novel addresses the impact of climate change on a rural Appalachian community, focusing on the arrival of displaced monarch butterflies and the protagonist's personal transformation.

Film: Visualizing Climate Change

Films have a unique ability to bring the realities of climate change to life, using visual storytelling to engage audiences emotionally and intellectually. Documentaries and fictional films alike play crucial roles in raising awareness and fostering dialogue.

Documentaries

"An Inconvenient Truth": Al Gore's landmark documentary brought climate change to the forefront of public consciousness, presenting scientific evidence and personal anecdotes to advocate for urgent action.

"Before the Flood": Produced and narrated by Leonardo DiCaprio, this documentary explores the impacts of climate change around the world and highlights the need for immediate global action.

Fictional Films

"The Day After Tomorrow": This blockbuster film dramatizes the effects of abrupt climate change, depicting a series of extreme weather events that lead to a new ice age. While scientifically exaggerated, it raises important questions about climate preparedness.

"Interstellar": Set in a future where climate change has rendered Earth nearly uninhabitable, this sci-fi film explores themes of exploration, survival, and the search for a new home for humanity.

Case Studies

"Chasing Ice": This documentary follows photographer James Balog's efforts to capture the melting of glaciers through time-lapse photography, providing powerful visual evidence of climate change.

"Snowpiercer": This dystopian film depicts a future where the remnants of humanity survive on a perpetually moving train after a failed climate engineering experiment. It explores themes of class struggle, survival, and environmental degradation.

Visual Arts: Documenting and Interpreting Change

Art as Witness

Visual artists play a critical role in documenting the effects of climate change and interpreting its significance. Through painting, photography, sculpture, and other media, artists can capture the beauty, fragility, and devastation of our changing world.

Photography

Environmental Documentation: Photographers like Sebastião Salgado and Edward Burtynsky document environmental destruction and human impact on the planet, creating compelling visual narratives that raise awareness and provoke reflection.

Climate Impact: Photographers such as Camille Seaman capture the stark beauty and vulnerability of polar ice, bringing the impacts of climate change in remote regions closer to the global audience.

Painting and Sculpture

Abstract Interpretation: Artists like Zaria Forman use pastel drawings to create large-scale, hyper-realistic images of melting glaciers and stormy seas, evoking an emotional response to climate change.

Installations: Environmental artist Andy Goldsworthy creates ephemeral sculptures using natural materials, highlighting the impermanence and fragility of the natural world.

Public Art and Activism

Murals and Street Art: Public art, including murals and street art, can engage communities and raise awareness about climate change. Artists like Banksy and Shepard Fairey use their work to comment on environmental issues and inspire action.

Eco-Art Projects: Eco-artists create works that not only raise awareness but also contribute to environmental restoration. Projects like Agnes Denes's "Wheatfield – A Confrontation" transform urban spaces into living artworks that promote sustainability.

Case Studies

Olafur Eliasson's "Ice Watch": Eliasson's installation involves placing large blocks of glacial ice in public spaces, where they slowly melt, serving as a powerful, tangible reminder of climate change.

Maya Lin's "What Is Missing?": This multimedia project documents species and ecosystems lost to climate change and habitat destruction, creating a poignant memorial that also educates and motivates viewers to protect what remains.

Performing Arts: Advocacy through Creativity

Theater and Performance Art

The performing arts, including theater, dance, and performance art, offer dynamic and immersive ways to explore and communicate the complexities of climate change. Performers can create powerful emotional experiences that inspire reflection and action.

Theater

Environmental Plays: Plays like "The Contingency Plan" by Steve Waters and "Greenland" by Moira Buffini, Matt Charman, Penelope Skinner, and Jack Thorne address climate change directly, exploring its scientific, political, and personal dimensions.

Interactive Performances: Interactive and immersive theater, such as "2071" by Duncan Macmillan and Chris Rapley, engages audiences in the science and ethics of climate change, encouraging active participation and reflection.

Dance and Movement

Dance Performances: Choreographers like Anna Halprin and Carolyn Carlson create dance pieces that explore themes of nature, environmental destruction, and renewal, using movement to express the urgency of climate action.

Site-Specific Works: Site-specific dance performances, such as those by the collective Eiko & Koma, take place in natural or urban environments impacted by climate change, creating a direct connection between the performance and its ecological context.

Music and Sound Art

Music and sound art can convey the emotional and sensory dimensions of climate change, from the sounds of melting ice to compositions that evoke the beauty and vulnerability of nature.

Compositions and Concerts

Climate-Inspired Music: Composers like John Luther Adams, known for works such as "Become Ocean," create music that reflects the natural world and its changing state, inspiring listeners to contemplate the environment.

Benefit Concerts: Benefit concerts and music festivals, such as Live Earth, bring together artists and audiences to raise awareness and funds for climate action, using the power of music to drive social change.

Case Studies

The Arctic Cycle: Playwright Chantal Bilodeau's initiative, The Arctic Cycle, involves a series of plays and projects that explore the impact of climate change on the Arctic and its inhabitants, fostering dialogue and action through storytelling.

"The Great Immensity" by The Civilians: This musical play by the investigative theater company The Civilians tackles themes of climate change and environmental science, using humor and drama to engage audiences.

Advocacy through Creativity

Artists and performers can also engage in direct advocacy, using their platforms and creative talents to support climate movements, influence policy, and inspire public action.

Collaborative Projects

Artivism: Artists collaborate with environmental organizations to create works that support climate campaigns, such as Greenpeace's collaborations with visual artists and musicians to produce impactful advocacy materials.

Educational Programs: Many artists develop educational programs and workshops that use creative expression to teach about climate change, empowering participants to become advocates in their own communities.

Case Studies

Extinction Rebellion's "Red Rebel Brigade": This performance art group, part of the climate activist movement Extinction Rebellion, uses striking visual imagery and silent, processional performances to draw attention to the climate crisis.

David Buckland's "Cape Farewell": The Cape Farewell project brings together artists, scientists, and communicators to inspire the creation of art that addresses climate change, fostering interdisciplinary collaboration and public engagement.

This chapter examines the diverse ways in which art and expression contribute to the climate crisis narrative. From literature and film to visual and performing arts, creative works not only document and interpret climate change but also advocate for action and inspire a deeper connection with the natural world. By leveraging the power of creativity, artists and performers play a vital role in shaping public perception and motivating collective efforts to address the climate emergency.

Chapter 14
Climate Change and Urban Identity

Urbanization and Environmental Footprints

The Rise of Urbanization

Urbanization, the process by which more people move to cities and urban areas, has been a defining trend of the 21st century. Currently, over half of the world's population lives in urban areas, a figure that is expected to rise significantly in the coming decades.

Drivers of Urbanization

Economic Opportunities: Cities often provide greater access to jobs, education, and healthcare, drawing people from rural areas seeking better livelihoods.

Infrastructure Development: Improvements in transportation, housing, and communication make urban living more attractive and accessible.

Environmental Footprints of Cities

While cities are centers of economic activity and innovation, they also have significant environmental impacts. Urban areas contribute disproportionately to climate change due to their high energy consumption, waste generation, and land use changes.

Carbon Emissions

Energy Use: Cities consume large amounts of energy for heating, cooling, transportation, and industrial activities, leading to high greenhouse gas emissions.

Transportation: The concentration of vehicles in urban areas contributes to air pollution and carbon emissions. Public transportation systems can mitigate some of these impacts, but many cities still rely heavily on private cars.

Waste Management

Solid Waste: Urban areas generate vast quantities of solid waste, much of which ends up in landfills or the ocean, contributing to pollution and greenhouse gas emissions.

Wastewater: The treatment and disposal of wastewater from urban areas can impact water quality and aquatic ecosystems.

Land Use and Biodiversity

Urban Sprawl: The expansion of urban areas often leads to the destruction of natural habitats and loss of biodiversity. This sprawl also increases the urban heat island effect, exacerbating climate impacts.

Green Spaces: While cities can support biodiversity through parks and green spaces, these areas are often limited and under threat from development.

Case Studies

New York City: As one of the largest cities in the world, New York City has significant environmental footprints but is also a leader in implementing green initiatives, such as expanding its public transportation system and increasing energy efficiency in buildings.

Beijing: Rapid urbanization in Beijing has led to severe air pollution and environmental degradation. The city is now investing in renewable energy and public transportation to reduce its environmental impact.

Resilient Cities: Redesigning for Sustainability

Principles of Urban Resilience

Urban resilience refers to the ability of cities to withstand and recover from environmental, economic, and social shocks. In the context of climate change, resilient cities are those that can adapt to and mitigate the impacts of climate change through sustainable design and planning.

Sustainable Infrastructure

Green Buildings: Incorporating energy-efficient technologies, green roofs, and sustainable materials in building design reduces energy consumption and greenhouse gas emissions.

Renewable Energy: Investing in renewable energy sources, such as solar and wind power, decreases reliance on fossil fuels and enhances energy security.

Transportation and Mobility

Public Transit: Developing efficient and accessible public transportation systems reduces traffic congestion and carbon emissions.

Active Transportation: Promoting walking and cycling through the creation of safe pedestrian and cycling infrastructure fosters healthier lifestyles and reduces environmental impacts.

Nature-Based Solutions

Nature-based solutions involve using natural processes and green infrastructure to address urban challenges, enhancing resilience and sustainability.

Urban Green Spaces

Parks and Gardens: Urban parks, community gardens, and green belts provide recreational spaces, improve air quality, and support biodiversity.

Green Roofs and Walls: These installations reduce the urban heat island effect, improve insulation, and manage stormwater runoff.

Water Management

Sustainable Drainage Systems: Implementing permeable pavements, rain gardens, and wetlands helps manage stormwater, reducing the risk of flooding and improving water quality.

Water Recycling: Reusing greywater and harvesting rainwater can reduce demand on freshwater resources and enhance water security.

Climate Adaptation and Mitigation Strategies

Heatwave Preparedness: Developing cooling centers, increasing tree canopy coverage, and implementing reflective surfaces can mitigate the impacts of heatwaves on urban populations.

Flood Resilience: Constructing levees, seawalls, and flood-resistant buildings, along with restoring natural floodplains, protects cities from rising sea levels and extreme weather events.

Case Studies

Copenhagen: Known for its ambitious climate policies, Copenhagen aims to become carbon neutral by 2025 through investments in renewable energy, green transportation, and sustainable urban design.

Singapore: Singapore's approach to urban resilience includes extensive green infrastructure, efficient public transportation, and innovative water management practices.

Community Identity: Urban Green Spaces and Social Cohesion

The Role of Green Spaces in Urban Life

Urban green spaces play a crucial role in shaping the identity and well-being of city dwellers. These spaces provide areas for recreation, relaxation, and social interaction, contributing to a sense of community and belonging.

Benefits of Green Spaces

Mental Health: Access to green spaces has been shown to reduce stress, improve mood, and enhance overall mental health.

Physical Health: Parks and recreational areas encourage physical activity, reducing the risk of chronic diseases and promoting healthier lifestyles.

Social Interaction: Green spaces serve as communal areas where people can meet, socialize, and build relationships, fostering social cohesion and community spirit.

Community-Led Green Initiatives

Community involvement in the creation and maintenance of green spaces can empower residents, promote environmental stewardship, and strengthen community bonds.

Urban Gardening and Agriculture

Community Gardens: These spaces allow residents to grow their own food, learn about sustainable agriculture, and connect with neighbors.

Urban Farms: Urban farming initiatives provide fresh, locally grown produce, reduce food miles, and promote food security.

Participatory Planning

Collaborative Design: Involving community members in the planning and design of green spaces ensures that these areas meet local needs and preferences.

Citizen Science: Engaging residents in monitoring and maintaining green spaces fosters a sense of ownership and responsibility for the environment.

Cultural and Historical Significance

Green spaces can also preserve and celebrate cultural and historical heritage, contributing to a city's unique identity and sense of place.

Heritage Parks

Preserving History: Parks that incorporate historical landmarks and cultural monuments help preserve local history and provide educational opportunities.

Cultural Events: Hosting cultural festivals, performances, and art installations in green spaces celebrates diversity and enriches community life.

Case Studies

High Line, New York City: This elevated linear park, built on a former railway track, has transformed an industrial relic into a vibrant public space, enhancing community identity and urban resilience.

Garden City Movement, Singapore: Singapore's vision of a "City in a Garden" integrates green spaces throughout the urban landscape, promoting biodiversity, enhancing quality of life, and strengthening community ties.

This chapter explores the complex relationship between climate change and urban identity, highlighting the environmental footprints of cities, strategies for building resilient urban areas, and the importance of green spaces in fostering community identity and social cohesion.

By reimagining urban living through sustainable practices and community engagement, cities can not only mitigate the impacts of climate change but also enhance the well-being and resilience of their residents.

Chapter 15
Rural and Agricultural Identities in Transition

Changing Landscapes: Agriculture and Climate

The Impact of Climate Change on Agriculture

Agriculture is highly vulnerable to the effects of climate change, which can disrupt traditional farming practices and alter landscapes. Rising temperatures, changing precipitation patterns, and extreme weather events pose significant challenges to farmers globally.

Temperature Changes

Crop Yield: Higher temperatures can affect crop growth cycles, reduce yields, and increase the prevalence of pests and diseases.

Livestock: Heat stress can impact livestock health and productivity, leading to reduced meat and dairy production.

Precipitation Patterns

Droughts: Prolonged droughts reduce water availability for irrigation, affecting crop and livestock production.

Floods: Increased rainfall and flooding can lead to soil erosion, nutrient loss, and damage to crops and infrastructure.

Extreme Weather Events

Storms and Hurricanes: Severe weather can destroy crops, livestock, and farming infrastructure, leading to significant economic losses.

Frost and Hail: Unexpected frost and hail can damage crops and reduce yields.

Shifting Agricultural Zones

Climate change is causing shifts in agricultural zones, forcing farmers to adapt their practices and crops to new conditions.

Crop Diversification

Alternative Crops: Farmers are exploring new crops that are more resilient to changing climate conditions, such as drought-resistant varieties.

Agroforestry: Integrating trees and shrubs into farming systems can enhance resilience, improve soil health, and provide additional income sources.

Precision Agriculture

Technology Use: Precision agriculture technologies, such as remote sensing and GPS, help farmers optimize resource use and improve crop management.

Data-Driven Decisions: Access to climate data and predictive models enables farmers to make informed decisions about planting, irrigation, and pest control.

Case Studies

California, USA: California's Central Valley, a major agricultural hub, faces challenges from prolonged droughts and water scarcity. Farmers are adopting water-efficient irrigation technologies and shifting to drought-tolerant crops.

Sahel Region, Africa: In the Sahel, farmers are practicing agroforestry and soil conservation techniques to combat desertification and improve agricultural productivity.

Farmers and Fishers: Adapting Traditional Livelihoods

Challenges Faced by Farmers

Farmers worldwide are grappling with the need to adapt their traditional practices to cope with the impacts of climate change.

Economic Pressures

Market Volatility: Climate-induced fluctuations in crop yields can lead to market instability and financial insecurity for farmers.

Cost of Adaptation: Implementing adaptive measures, such as new technologies and infrastructure, can be financially burdensome for small-scale farmers.

Knowledge and Resources

Access to Information: Farmers need access to climate information, training, and resources to effectively adapt their practices.

Extension Services: Agricultural extension services play a critical role in disseminating knowledge and supporting farmers in their adaptation efforts.

Adaptation Strategies for Farmers

Farmers are employing a range of strategies to adapt to the changing climate and sustain their livelihoods.

Sustainable Farming Practices

Conservation Agriculture: Practices such as minimal soil disturbance, crop rotation, and cover cropping improve soil health and resilience.

Organic Farming: Organic farming methods reduce reliance on chemical inputs, enhance biodiversity, and build resilience to climate shocks.

Water Management

Efficient Irrigation: Drip and sprinkler irrigation systems reduce water use and increase water use efficiency.

Rainwater Harvesting: Collecting and storing rainwater provides an additional water source for irrigation during dry periods.

Challenges Faced by Fishers

Climate change also significantly impacts fisheries and coastal communities, threatening their traditional livelihoods.

Ocean Changes

Rising Temperatures: Warming oceans affect fish distribution, breeding cycles, and growth rates, leading to shifts in fish populations.

Acidification: Ocean acidification, resulting from increased CO_2 absorption, impacts shellfish and coral reefs, affecting marine biodiversity.

Extreme Weather and Sea-Level Rise

Storms: More frequent and severe storms damage fishing infrastructure, boats, and coastal habitats.

Sea-Level Rise: Rising sea levels threaten coastal communities and fishing grounds, leading to displacement and loss of livelihoods.

Adaptation Strategies for Fishers

Fishers are adopting various strategies to adapt to the impacts of climate change and sustain their livelihoods.

Sustainable Fishing Practices

Ecosystem-Based Management: Managing fisheries with an ecosystem approach ensures the sustainability of fish stocks and marine habitats.

Selective Gear: Using selective fishing gear reduces bycatch and minimizes damage to marine ecosystems.

Aquaculture

Fish Farming: Aquaculture provides an alternative livelihood for fishers, reducing pressure on wild fish populations and enhancing food security.

Integrated Systems: Integrating aquaculture with agriculture, such as rice-fish farming, optimizes resource use and increases resilience.

Case Studies

Bangladesh: Coastal fishers in Bangladesh are adopting sustainable fishing practices and exploring alternative livelihoods, such as aquaculture, to cope with the impacts of climate change.

Norway: Norwegian fishers are utilizing technology and sustainable management practices to adapt to changing fish populations and maintain their livelihoods.

Rural Community Resilience: Collective Adaptation

Building Community Resilience

Rural communities are leveraging collective efforts and social cohesion to build resilience to climate change. Community-based approaches enhance adaptive capacity and ensure that adaptation measures are inclusive and sustainable.

Community-Led Initiatives

Local Knowledge: Incorporating traditional knowledge and practices into adaptation strategies enhances their relevance and effectiveness.

Participatory Planning: Engaging community members in the planning and implementation of adaptation projects ensures that they address local needs and priorities.

Social Networks and Support Systems

Strong social networks and support systems are critical for building community resilience to climate change.

Mutual Support

Cooperatives: Farmer and fisher cooperatives provide mutual support, access to resources, and collective bargaining power.

Community Savings Groups: Savings groups help community members pool resources and provide financial support during climate shocks.

Knowledge Sharing

Peer Learning: Peer-to-peer learning and knowledge exchange enable communities to share successful adaptation practices and innovations.

Extension Services: Strengthening agricultural and fisheries extension services ensures that communities have access to up-to-date information and support.

Diversifying Livelihoods

Diversifying livelihoods reduces dependence on a single income source and enhances resilience to climate impacts.

Non-Farm Activities

Handicrafts and Small Enterprises: Developing handicrafts and small businesses provides additional income sources for rural households.

Ecotourism: Promoting ecotourism leverages natural and cultural heritage, creating sustainable income opportunities.

Education and Training

Skills Development: Providing education and training in alternative livelihoods and climate-resilient practices empowers community members to adapt and thrive.

Youth Engagement: Engaging youth in climate adaptation efforts ensures the sustainability of community resilience strategies.

Case Studies

Nepal: Rural communities in Nepal are practicing community forestry and agroforestry to enhance resilience, improve livelihoods, and protect natural resources.

Pacific Islands: Pacific Island communities are implementing traditional knowledge and modern technologies to adapt to rising sea levels and changing marine ecosystems.

This chapter explores the profound impact of climate change on rural and agricultural identities, highlighting the challenges faced by farmers and fishers and the strategies they are adopting to adapt.

Through collective adaptation and diversification of livelihoods, rural communities are building resilience and sustaining their way of life in the face of a changing climate. By leveraging traditional knowledge, social networks, and innovative practices, these communities are navigating the transition and forging a path towards a more resilient future.

Chapter 16
The Role of Education in Shaping Climate Identity

Curriculum Changes: Integrating Climate Education

The Importance of Climate Education

Climate education plays a crucial role in shaping the next generation's understanding of and response to climate change. By integrating climate topics into the curriculum, educational institutions can empower students with the knowledge and skills needed to address environmental challenges.

Early Education

Foundational Knowledge: Introducing climate concepts at an early age builds a foundation for lifelong environmental awareness and responsibility.

Engagement through Activities: Hands-on activities, such as school gardens and recycling projects, engage young learners and make climate concepts tangible.

Secondary Education

Interdisciplinary Approach: Incorporating climate education across subjects, including science, geography, and social studies, provides a comprehensive understanding of climate change.

Critical Analysis: Encouraging students to analyze climate data, debate policies, and explore solutions fosters critical thinking and problem-solving skills.

Higher Education

Specialized Courses: Offering courses and programs focused on climate science, sustainability, and environmental policy prepares students for careers in climate-related fields.

Research Opportunities: Providing opportunities for students to engage in climate research and projects enhances their practical skills and contributes to scientific knowledge.

Integrating Climate Education into the Curriculum

Effective integration of climate education requires a holistic and collaborative approach, involving educators, policymakers, and communities.

Curriculum Development

Standards and Frameworks: Developing national and international standards for climate education ensures consistency and comprehensiveness across educational institutions.

Teacher Training: Equipping teachers with the knowledge and resources to effectively teach climate topics is essential for successful implementation.

Innovative Teaching Methods

Experiential Learning: Field trips, outdoor education, and service-learning projects provide experiential learning opportunities that deepen students' understanding of climate issues.

Technology Integration: Utilizing digital tools and platforms, such as virtual simulations and interactive maps, enhances climate education and engagement.

Case Studies

Finland: Finland's education system integrates climate education across all levels, emphasizing sustainability, outdoor learning, and interdisciplinary teaching methods.

Costa Rica: Costa Rican schools incorporate environmental education into their curricula, focusing on biodiversity, conservation, and sustainable practices.

Critical Thinking: Empowering Informed Citizens

The Role of Critical Thinking in Climate Education

Critical thinking is essential for understanding the complexities of climate change and making informed decisions. Educating students to think critically about climate issues empowers them to become engaged and informed citizens.

Analyzing Climate Data

Data Literacy: Teaching students how to interpret and analyze climate data, such as temperature records and carbon emissions, develops their data literacy and analytical skills.

Scientific Inquiry: Encouraging scientific inquiry and experimentation helps students understand the processes and impacts of climate change.

Evaluating Sources and Information

Media Literacy: Educating students on how to critically evaluate media sources and recognize misinformation is crucial in an era of widespread misinformation about climate change.

Debates and Discussions: Facilitating debates and discussions on climate policies and solutions promotes critical thinking and diverse perspectives.

Empowering Action through Knowledge

Knowledge alone is not enough; empowering students to take action is a key goal of climate education.

Advocacy and Activism

Youth Movements: Encouraging students to participate in youth climate movements, such as Fridays for Future, fosters a sense of agency and collective action.

Policy Engagement: Educating students on how to engage with policymakers and advocate for climate policies empowers them to influence decision-making processes.

Sustainable Practices

Behavioral Change: Teaching students about sustainable practices, such as energy conservation and waste reduction, encourages them to adopt environmentally friendly behaviors.

Community Projects: Involving students in community projects, such as tree planting and clean-up drives, connects them with local environmental initiatives.

Case Studies

Germany: German schools emphasize critical thinking and environmental education, encouraging students to engage in climate activism and sustainable practices.

New Zealand: New Zealand's education system integrates critical thinking and sustainability, promoting environmental stewardship and active citizenship.

Lifelong Learning: Community and Global Initiatives

The Need for Lifelong Climate Education

Climate education should not be limited to formal schooling; it is a lifelong process that involves continuous learning and adaptation. Community and global initiatives play a crucial role in fostering lifelong learning and climate awareness.

Community-Based Education

Workshops and Seminars: Offering workshops and seminars on climate topics in community centers and libraries provides accessible education for all age groups.

Citizen Science: Engaging community members in citizen science projects, such as monitoring local ecosystems, enhances their understanding and involvement in climate issues.

Global Educational Initiatives

International Collaborations: Collaborating with international organizations and educational institutions promotes the exchange of knowledge and best practices in climate education.

Online Platforms: Utilizing online platforms and courses, such as MOOCs (Massive Open Online Courses), makes climate education accessible to a global audience.

Role of NGOs and Non-Profits

Non-governmental organizations (NGOs) and non-profits play a vital role in providing climate education and fostering lifelong learning.

Outreach Programs

School Partnerships: Partnering with schools to deliver climate education programs and resources enhances the reach and impact of educational initiatives.

Public Campaigns: Running public awareness campaigns on climate issues informs and engages the broader community.

Educational Resources

Toolkits and Guides: Developing educational toolkits and guides for teachers, parents, and community leaders supports climate education efforts.

Interactive Learning: Creating interactive learning materials, such as games and apps, makes climate education engaging and accessible.

Lifelong Learning for Professionals

Continuing education for professionals in various fields is essential for addressing climate challenges and integrating sustainability into all sectors.

Professional Development

Training Programs: Offering training programs and certifications in climate science, sustainability, and environmental management enhances professional skills and knowledge.

Industry Partnerships: Collaborating with industries to develop sustainable practices and technologies fosters innovation and climate resilience.

Case Studies

Australia: Australia's community-based climate education initiatives, such as Landcare and Coastcare, engage local communities in environmental stewardship and lifelong learning.

Sweden: Sweden's adult education programs, including study circles and folk high schools, promote lifelong learning and climate awareness among citizens.

This chapter explores the critical role of education in shaping climate identity, highlighting the importance of integrating climate education into curricula, fostering critical thinking, and promoting lifelong learning. By empowering individuals with knowledge and skills, education can drive meaningful action and resilience in the face of climate change. Through community and global initiatives, continuous learning, and professional development, we can build a more informed and engaged society, capable of addressing the challenges of a changing climate.

Chapter 17
Policy, Governance, and Identity

Climate Policy: Local, National, and Global Frameworks

Local Climate Policies

Local governments play a crucial role in implementing climate policies that directly impact communities. These policies address specific environmental challenges and harness local strengths and resources.

Urban Planning and Zoning

Green Infrastructure: Cities are incorporating green roofs, parks, and urban forests to mitigate the urban heat island effect and improve air quality.

Public Transportation: Investments in public transportation systems reduce traffic congestion and lower carbon emissions, fostering more sustainable urban environments.

Community Initiatives

Energy Efficiency Programs: Local governments promote energy efficiency in homes and businesses through incentives for retrofitting and the use of energy-efficient appliances.

Waste Management: Programs aimed at reducing, reusing, and recycling waste help minimize landfill use and reduce greenhouse gas emissions.

National Climate Policies

National governments establish broader frameworks and regulations that drive climate action and set the stage for local and international efforts.

Legislation and Regulation

Carbon Pricing: Implementing carbon taxes or cap-and-trade systems incentivizes businesses to reduce their carbon footprints.

Renewable Energy Mandates: National policies that mandate a certain percentage of energy to come from renewable sources accelerate the transition away from fossil fuels.

National Adaptation Strategies

Disaster Preparedness: Developing strategies to prepare for and respond to climate-related disasters, such as floods, hurricanes, and wildfires, enhances national resilience.

Research and Innovation: Funding research into climate science and green technologies fosters innovation and supports the development of sustainable solutions.

Global Climate Frameworks

Global cooperation is essential to address the transboundary nature of climate change. International agreements and organizations provide platforms for coordinated action.

International Agreements

Paris Agreement: This landmark agreement aims to limit global warming to well below 2°C above pre-industrial levels, with efforts to limit the increase to 1.5°C. Countries set nationally determined contributions (NDCs) to achieve these goals.

Kyoto Protocol: Preceding the Paris Agreement, the Kyoto Protocol established binding emission reduction targets for developed countries.

Global Organizations

United Nations Framework Convention on Climate Change (UNFCCC): The UNFCCC facilitates international climate negotiations and supports countries in meeting their climate commitments.

Intergovernmental Panel on Climate Change (IPCC): The IPCC provides scientific assessments on climate change, informing policy decisions and public understanding.

Case Studies

Germany: Germany's Energiewende (energy transition) policy promotes renewable energy, energy efficiency, and sustainable development, making it a global leader in climate action.

Costa Rica: Costa Rica has achieved significant success in reforestation and renewable energy, aiming to become a carbon-neutral country by 2050.

Governance and Trust: Public Perception and Engagement

The Role of Governance in Climate Action

Effective governance is crucial for implementing climate policies and ensuring public trust and engagement. Transparent, accountable, and inclusive governance fosters public support and participation in climate action.

Transparency and Accountability

Open Data: Governments that provide access to climate data and policy information enable public scrutiny and foster trust.

Accountability Mechanisms: Establishing independent oversight bodies and mechanisms for monitoring and reporting progress ensures that governments are held accountable for their climate commitments.

Public Participation

Consultation Processes: Engaging citizens in the policy-making process through consultations, public hearings, and participatory budgeting increases public buy-in and the legitimacy of climate policies.

Community Involvement: Encouraging community-led initiatives and local climate action plans empowers residents to take ownership of their climate resilience efforts.

Building Trust in Climate Governance

Public trust is essential for the successful implementation of climate policies. Trust is built through consistent, credible, and equitable actions by governments and institutions.

Consistency and Credibility

Policy Coherence: Consistent policies across different levels of government and sectors build credibility and ensure effective implementation.

Evidence-Based Decision Making: Basing policies on scientific evidence and expert advice reinforces public confidence in climate governance.

Equity and Justice

Inclusive Policies: Climate policies that address the needs of vulnerable and marginalized communities ensure that the benefits and burdens of climate action are equitably distributed.

Climate Justice: Promoting climate justice involves recognizing and addressing the disproportionate impacts of climate change on low-income and marginalized communities.

Public Perception and Engagement

Public perception of climate change and government actions significantly influences the success of climate policies. Effective communication and engagement strategies are key to shaping positive perceptions and mobilizing collective action.

Communication Strategies

Clear Messaging: Communicating the science and impacts of climate change in clear, relatable terms helps build public understanding and support.

Positive Framing: Highlighting the co-benefits of climate action, such as improved health, economic opportunities, and enhanced quality of life, motivates public engagement.

Education and Awareness

Climate Literacy: Enhancing climate literacy through education campaigns and public outreach programs empowers citizens to make informed decisions and advocate for climate action.

Media and Social Media: Leveraging traditional and social media platforms to disseminate information and engage with diverse audiences amplifies the reach and impact of climate messages.

Case Studies

Sweden: Sweden's government prioritizes transparency and public participation in its climate policies, fostering high levels of public trust and engagement.

Bhutan: Bhutan's commitment to maintaining carbon neutrality and its emphasis on Gross National Happiness demonstrate a holistic approach to climate governance that integrates well-being and sustainability.

Identity Politics: Advocacy and Representation

The Role of Identity in Climate Advocacy

Identity politics plays a significant role in climate advocacy, as individuals and groups mobilize around shared identities and values to advocate for climate action.

Grassroots Movements

Youth Movements: Youth-led movements, such as Fridays for Future, mobilize young people around their collective identity and future stakes in climate action.

Indigenous Movements: Indigenous communities advocate for climate action based on their deep connection to the land and traditional ecological knowledge.

Environmental Justice

Climate Justice: Advocates for climate justice emphasize the need to address the disproportionate impacts of climate change on marginalized communities and promote equitable solutions.

Intersectionality: Recognizing the intersection of climate issues with race, gender, and socioeconomic status highlights the diverse experiences and needs of different communities.

Representation in Climate Policy

Inclusive representation in climate policy ensures that diverse voices and perspectives are considered, leading to more equitable and effective solutions.

Political Representation

Diverse Leadership: Increasing the representation of women, indigenous peoples, and marginalized groups in political and decision-making processes enhances the inclusivity of climate policies.

Advocacy Groups: Environmental NGOs and advocacy groups play a critical role in representing the interests of various communities and influencing policy decisions.

Participatory Governance

Citizen Assemblies: Establishing citizen assemblies or climate councils that include representatives from diverse backgrounds fosters inclusive decision-making.

Community-Led Initiatives: Supporting community-led climate initiatives empowers local groups to take action and ensures that policies reflect their needs and priorities.

Advocacy Strategies

Effective advocacy strategies leverage identity and representation to mobilize support and drive climate action.

Coalition Building

Alliances and Networks: Building alliances among diverse groups, such as environmental organizations, social justice groups, and labor unions, strengthens advocacy efforts and broadens support.

Intersectional Advocacy: Advocating for policies that address multiple intersecting issues, such as climate justice and economic equity, creates more comprehensive and impactful solutions.

Storytelling and Narrative Building

Personal Stories: Sharing personal stories and experiences of climate impacts humanizes the issue and resonates with broader audiences.

Cultural Narratives: Leveraging cultural narratives and traditions to frame climate issues in relatable terms fosters deeper engagement and connection.

Case Studies

United States: The Sunrise Movement, a youth-led climate advocacy group, effectively uses identity politics and storytelling to advocate for a Green New Deal and comprehensive climate action.

Canada: Indigenous-led movements in Canada, such as Idle No More, advocate for environmental justice and the protection of indigenous lands and rights in the face of climate change.

This chapter explores the complex interplay between policy, governance, and identity in the context of climate change. By examining local, national, and global climate policies, the role of governance in building trust and public engagement, and the impact of identity politics on advocacy and representation, this chapter highlights the importance of inclusive and equitable climate action.

Through effective governance, diverse representation, and empowered advocacy, societies can navigate the challenges of climate change and build resilient, sustainable futures.

Chapter 18
Ethical and Philosophical Dimensions of Climate Change

Moral Imperatives: Responsibility and Stewardship

Responsibility to Future Generations

Climate change poses a unique moral challenge due to its long-term impacts and the need for intergenerational justice. The ethical responsibility to future generations demands that we take action now to prevent harm and ensure a livable planet for those who come after us.

Intergenerational Justice

Moral Duty: We have a moral duty to ensure that future generations inherit a world that is not irreparably damaged by our actions. This involves reducing carbon emissions, preserving natural resources, and maintaining biodiversity.

Precautionary Principle: Adopting the precautionary principle means taking preventive action in the face of uncertainty to avoid severe or irreversible damage to the environment and human health.

Legacy and Heritage

Cultural and Natural Heritage: Protecting cultural sites, traditions, and natural landscapes from the effects of climate change preserves the legacy for future generations and maintains the diversity of human experience and knowledge.

Education and Awareness: Educating young people about climate change and sustainability instills a sense of responsibility and empowers them to take action in their own lives and communities.

Stewardship of the Earth

The concept of stewardship emphasizes our responsibility to care for the Earth and its ecosystems, recognizing that human well-being is deeply interconnected with the health of the planet.

Environmental Stewardship

Sustainable Resource Use: Responsible stewardship involves using natural resources in a way that meets current needs without compromising the ability of future generations to meet their own needs. This includes practices such as sustainable agriculture, forestry, and fishing.

Conservation Efforts: Protecting endangered species and preserving habitats through conservation efforts ensures the survival of biodiversity, which is vital for ecosystem health and resilience.

Ethical Consumption

Consumer Choices: Individuals can exercise stewardship by making ethical consumer choices, such as purchasing sustainably sourced products, reducing waste, and supporting companies with environmentally responsible practices.

Corporate Responsibility:

Corporations have a responsibility to minimize their environmental impact, adopt sustainable practices, and contribute to global efforts to combat climate change.

Case Studies

Norway: Norway's approach to environmental stewardship includes the preservation of natural landscapes, investment in renewable energy, and a commitment to sustainable development.

New Zealand: New Zealand's legal recognition of the Whanganui River as a living entity with rights reflects an innovative approach to environmental stewardship and indigenous values.

Philosophical Debates: Human-Nature Relationships

Anthropocentrism vs. Ecocentrism

The philosophical debate between anthropocentrism and ecocentrism centers on the value and significance of nature relative to human interests.

Anthropocentrism

Human-Centered View: Anthropocentrism places human beings at the center of moral consideration, valuing nature primarily for its utility to humans. This perspective often justifies exploiting natural resources for economic growth and development.

Ethical Implications: Critics argue that anthropocentrism leads to environmental degradation and fails to recognize the intrinsic value of nature. It prioritizes short-term human benefits over long-term environmental health.

Ecocentrism

Nature-Centered View: Ecocentrism recognizes the intrinsic value of all living beings and ecosystems, independent of their utility to humans. It advocates for the protection of nature for its own sake.

Ethical Implications: Ecocentrism promotes a more holistic approach to environmental ethics, emphasizing the interconnectedness of all life and the importance of maintaining the balance and integrity of ecosystems.

Deep Ecology and the Rights of Nature

Deep ecology and the rights of nature movement challenge conventional views of human dominance over nature and advocate for a more egalitarian relationship with the natural world.

Deep Ecology

Philosophical Foundations: Deep ecology, founded by Arne Naess, calls for a profound shift in human consciousness and values, recognizing the intrinsic worth of all living beings and ecosystems.

Practical Applications: Deep ecology encourages sustainable living, conservation efforts, and policies that respect the inherent value of nature. It promotes simplicity, mindfulness, and a deep connection to the natural world.

Rights of Nature

Legal Recognition: The rights of nature movement seeks to grant legal personhood and rights to natural entities, such as rivers, forests, and mountains. This legal framework recognizes nature as a subject with rights, rather than an object for human exploitation.

Ethical Implications: Recognizing the rights of nature shifts the ethical and legal paradigm, emphasizing the need to protect and preserve natural systems for their own sake and for the well-being of future generations.

Case Studies

Ecuador: Ecuador's constitution recognizes the rights of nature, granting ecosystems the right to exist, persist, and regenerate. This legal framework has been used to challenge environmentally harmful activities and promote sustainable development.

India: In India, the Ganges and Yamuna rivers have been granted legal personhood, acknowledging their cultural and ecological significance and providing a basis for their protection and restoration.

Ethical Frameworks: Guiding Climate Action

Utilitarianism and Climate Ethics

Utilitarianism, which advocates for actions that maximize overall happiness and well-being, provides a compelling ethical framework for addressing climate change.

Cost-Benefit Analysis

Balancing Benefits and Harms: Utilitarianism supports policies and actions that balance the benefits of economic development with the need to minimize environmental harm and mitigate climate change.

Global Perspective: Utilitarian ethics emphasize the importance of considering the well-being of all people, including those in developing countries who are disproportionately affected by climate change.

Ethical Trade-offs

Mitigation vs. Adaptation: Utilitarianism helps navigate the ethical trade-offs between investing in mitigation efforts to reduce future climate impacts and adaptation measures to protect vulnerable communities in the present.

Technological Solutions: Evaluating the potential benefits and risks of technological solutions, such as geoengineering and renewable energy technologies, through a utilitarian lens ensures that they contribute to overall well-being.

Deontological Ethics and Climate Duties

Deontological ethics, which focus on duties and principles, provide a robust framework for understanding our moral obligations to address climate change.

Duty to Prevent Harm

Moral Imperatives: Deontological ethics emphasize the duty to prevent harm and protect vulnerable populations from the adverse effects of climate change. This includes reducing greenhouse gas emissions and supporting climate adaptation efforts.

Rights-Based Approach: Recognizing the rights of individuals and communities to a healthy environment and a stable climate reinforces the moral obligation to take climate action.

Ethical Principles

Principle of Justice: Deontological ethics highlight the importance of justice and fairness in climate policy, ensuring that the burdens and benefits of climate action are equitably distributed.

Principle of Responsibility: Acknowledging the responsibility of individuals, corporations, and governments to mitigate climate change and support sustainable practices reflects deontological commitments to ethical conduct.

Virtue Ethics and Environmental Stewardship

Virtue ethics, which focus on the development of moral character and virtues, offer valuable insights into fostering environmental stewardship and ethical behavior in the face of climate change.

Cultivating Environmental Virtues

Virtues of Care and Respect: Virtue ethics emphasize the cultivation of care, respect, and compassion for the natural world, fostering a deep sense of responsibility and connection to the environment.

Sustainability as a Virtue: Embracing sustainability as a virtue involves developing habits and practices that support long-term ecological health and balance.

Role of Moral Exemplars

Leadership and Role Models: Moral exemplars, such as environmental activists and leaders who demonstrate virtuous behavior and commitment to sustainability, inspire others to adopt ethical practices and take action on climate change.

Community Engagement:

Encouraging community engagement and collective action reinforces the development of environmental virtues and promotes a culture of sustainability.

Case Studies

Bhutan:

Bhutan's commitment to Gross National Happiness, which prioritizes well-being and environmental sustainability over economic growth, reflects a holistic approach to ethical governance and climate action.

Scandinavia:

Scandinavian countries, such as Sweden and Denmark, exemplify the integration of ethical principles and environmental stewardship in their policies and practices, promoting renewable energy, sustainable living, and social equity.

This chapter delves into the ethical and philosophical dimensions of climate change, exploring moral imperatives, philosophical debates, and ethical frameworks that guide climate action. By examining the responsibility to future generations, the concept of stewardship, and various ethical perspectives, this chapter highlights the importance of integrating ethical considerations into our responses to climate change. Through a deeper understanding of these ethical and philosophical dimensions, we can develop more just, sustainable, and effective approaches to addressing the global climate crisis.

Chapter 19
Imagining the Future
Scenarios and Identity

Possible Futures: Optimistic, Pessimistic, and Realistic

Optimistic Futures: Achieving Sustainable Societies

In an optimistic future, humanity successfully addresses climate change through comprehensive global cooperation, technological innovation, and profound shifts in societal values and behaviors. This scenario envisions a world where sustainable practices become the norm, and the impacts of climate change are mitigated effectively.

Technological Advancements

Renewable Energy: Widespread adoption of renewable energy sources such as solar, wind, and hydroelectric power drastically reduces carbon emissions and dependence on fossil fuels.

Carbon Capture: Advances in carbon capture and storage technologies effectively remove excess carbon dioxide from the atmosphere, helping to stabilize global temperatures.

Social and Economic Transformations

Green Economies: Economies worldwide transition to green practices, prioritizing sustainable development, circular economies, and low-carbon industries.

Global Equity: International efforts to address climate justice result in equitable resource distribution and enhanced resilience for vulnerable populations.

Environmental Restoration

Reforestation and Conservation: Large-scale reforestation and conservation initiatives restore ecosystems, enhance biodiversity, and provide natural carbon sinks.

Sustainable Agriculture: Innovations in sustainable agriculture ensure food security while minimizing environmental impact through practices such as agroforestry, permaculture, and regenerative farming.

Pessimistic Futures: The Consequences of Inaction

A pessimistic future scenario highlights the dire consequences of insufficient action on climate change, characterized by severe environmental degradation, social upheaval, and widespread suffering.

Environmental Catastrophes

Extreme Weather Events: Increased frequency and intensity of hurricanes, droughts, floods, and heatwaves lead to devastating impacts on communities, infrastructure, and ecosystems.

Biodiversity Loss: Massive species extinctions and ecosystem collapses result from habitat destruction, pollution, and climate-induced changes.

Social and Economic Collapse

Mass Displacement: Rising sea levels, desertification, and resource scarcity force millions to become climate refugees, leading to humanitarian crises and conflicts over dwindling resources.

Economic Instability: Global economies suffer from the impacts of climate change, including agricultural failures, infrastructure damage, and increased costs for disaster response and recovery.

Health Crises

Public Health Threats: Climate change exacerbates health issues such as heat-related illnesses, respiratory diseases, and the spread of infectious diseases, overwhelming healthcare systems and reducing life expectancy.

Realistic Futures: Navigating Complex Challenges

A realistic future acknowledges the complexities and uncertainties of addressing climate change, combining elements of both optimistic and pessimistic scenarios. It emphasizes incremental progress, adaptive strategies, and ongoing efforts to balance environmental, social, and economic needs.

Incremental Progress

Policy and Governance: Gradual implementation of climate policies and international agreements leads to steady reductions in greenhouse gas emissions and improvements in sustainability practices.

Technological Innovations: Continued advancements in green technologies and renewable energy contribute to mitigating climate impacts, though challenges in scalability and adoption remain.

Adaptive Strategies

Resilience Building: Communities and governments focus on building resilience to climate impacts through adaptive infrastructure, disaster preparedness, and social safety nets.

Collaborative Efforts: Multisectoral collaboration among governments, businesses, and civil society fosters innovative solutions and shared responsibility for climate action.

Ongoing Challenges

Inequities and Justice: Despite progress, significant disparities in climate vulnerability and resilience persist, requiring ongoing efforts to address climate justice and support marginalized communities.

Environmental Degradation: Some degree of environmental degradation and biodiversity loss continues, necessitating continuous efforts in conservation and restoration.

Case Studies

Denmark: Denmark's transition to a low-carbon economy through renewable energy, green technologies, and sustainable practices demonstrates an optimistic path forward.

Bangladesh: Bangladesh's adaptive strategies, including community-based disaster risk reduction and climate-resilient infrastructure, illustrate realistic approaches to navigating climate challenges.

Adapting Identities: Personal and Collective Narratives

Personal Narratives: Redefining Self in a Changing World

As individuals face the realities of climate change, personal identities evolve to incorporate new values, responsibilities, and aspirations. Adapting personal narratives involves embracing sustainability, resilience, and a deep connection to the environment.

Sustainable Living

Lifestyle Changes: Adopting sustainable lifestyles, such as reducing consumption, minimizing waste, and supporting eco-friendly products, becomes a core aspect of personal identity.

Mindful Consumption: Individuals prioritize mindful consumption, making choices that align with environmental values and reduce their ecological footprint.

Environmental Stewardship

Active Participation: Engaging in environmental activism, conservation efforts, and community initiatives fosters a sense of purpose and connection to the natural world.

Education and Advocacy: Educating oneself and others about climate change and advocating for policy changes and sustainable practices reinforce a commitment to environmental stewardship.

Collective Narratives: Building Shared Visions

Communities and societies collectively adapt their identities to navigate the challenges and opportunities presented by climate change. Shared narratives of resilience, cooperation, and innovation shape collective identities and drive collective action.

Community Resilience

Collective Action: Communities come together to implement local solutions, such as community gardens, renewable energy projects, and disaster preparedness plans, enhancing collective resilience.

Cultural Adaptation: Cultural traditions and practices evolve to reflect changing environmental realities, preserving heritage while fostering adaptability and innovation.

Global Solidarity

Transnational Movements: Global movements, such as Fridays for Future and Extinction Rebellion, unite people across borders in the fight against climate change, creating a sense of global solidarity and shared identity.

Intercultural Exchange: Collaborative efforts and intercultural exchange promote mutual understanding and cooperation, enriching collective narratives and fostering a global sense of responsibility.

Case Studies

Japan: Japan's community-based disaster preparedness and cultural emphasis on harmony with nature illustrate the adaptation of collective narratives to enhance resilience.

The Netherlands: The Netherlands' innovative approaches to flood management and sustainable urban planning reflect a collective identity centered on resilience and environmental stewardship.

Building Hope: Visionary Leadership and Innovation

Visionary Leadership: Inspiring Change

Visionary leaders play a crucial role in inspiring and mobilizing collective action on climate change. Their ability to articulate compelling visions, demonstrate ethical leadership, and foster collaboration drives progress and instills hope.

Ethical Leadership

Commitment to Sustainability: Leaders who prioritize sustainability and environmental ethics set a powerful example, inspiring others to follow suit and adopt similar values.

Transparent Governance: Ethical leaders ensure transparency, accountability, and inclusivity in climate governance, building public trust and fostering a culture of shared responsibility.

Mobilizing Movements

Grassroots Leadership: Grassroots leaders and activists galvanize communities, amplifying marginalized voices and driving bottom-up change.

Institutional Leadership: Leaders within institutions, such as governments, businesses, and NGOs, leverage their positions to implement sustainable practices and advocate for systemic change.

Innovation: Catalyzing Transformative Solutions

Innovation is key to addressing the multifaceted challenges of climate change. Technological, social, and economic innovations catalyze transformative solutions and pave the way for a sustainable future.

Technological Innovation

Clean Energy Technologies: Innovations in renewable energy, energy storage, and smart grids enable a transition to low-carbon energy systems.

Sustainable Agriculture: Advances in sustainable agriculture practices, such as precision farming, vertical farming, and agroecology, enhance food security and reduce environmental impact.

Social Innovation

Community Initiatives: Innovative community initiatives, such as cooperative housing, shared mobility, and urban farming, promote sustainability and strengthen social bonds.

Behavioral Change: Initiatives that encourage behavioral change, such as public awareness campaigns and incentive programs, foster sustainable lifestyles and consumption patterns.

Visionary Projects

Green New Deal: The Green New Deal represents a visionary project that integrates climate action with social and economic justice, aiming to create a sustainable and equitable future.

Circular Economy Initiatives: Circular economy initiatives promote resource efficiency, waste reduction, and the creation of closed-loop systems, transforming production and consumption patterns.

Case Studies

Iceland: Iceland's transition to renewable energy and its innovative approaches to geothermal and hydropower exemplify visionary leadership and technological innovation.

Singapore: Singapore's efforts in urban sustainability, including green building standards, water management, and biodiversity conservation, highlight the power of innovation and visionary governance.

This chapter explores the imaginative potential of different future scenarios, the adaptation of personal and collective identities, and the role of visionary leadership and innovation in building hope. By examining optimistic, pessimistic, and realistic futures, the chapter highlights the importance of resilience, adaptability, and collaboration in navigating the challenges of climate change.

Through the evolution of personal and collective narratives, and the catalytic power of visionary leadership and innovation, societies can envision and work towards a sustainable and hopeful future.

Chapter 20
Conclusion
Integrating Knowledge and Action

Synthesis: Understanding the Interconnections

Holistic Understanding of Climate Change

Throughout this book, we have explored the multifaceted nature of climate change and its profound impacts on human identity. Climate change is not just an environmental issue; it is deeply intertwined with cultural, psychological, economic, and social dimensions.

Interconnected Impacts

Environmental and Social Systems: Climate change affects natural ecosystems and human societies in complex and interconnected ways. Understanding these interconnections is crucial for developing comprehensive solutions.

Cross-Disciplinary Approaches: Addressing climate change requires knowledge from various fields, including science, economics, psychology, and the humanities. A holistic approach that integrates these perspectives enhances our ability to respond effectively.

Identity and Adaptation

Evolving Identities: As individuals and communities confront climate change, their identities evolve. This evolution involves adapting values, lifestyles, and cultural practices to align with sustainability and resilience.

Resilience and Agency: Building resilience involves not only physical adaptations but also psychological and cultural shifts that empower individuals and communities to navigate uncertainties and challenges.

Systems Thinking

Complex Systems: Climate change is a product of complex, dynamic systems that include atmospheric, ecological, and social components. Systems thinking helps us understand how these components interact and influence each other.

Feedback Loops: Recognizing feedback loops, where changes in one part of the system influence others, is essential for anticipating and mitigating the cascading effects of climate change.

Case Studies

Pacific Island Nations: The Pacific Island nations' holistic approach to climate resilience, integrating traditional knowledge with modern science, exemplifies the power of cross-disciplinary and interconnected strategies.

Costa Rica: Costa Rica's commitment to renewable energy and conservation highlights the benefits of integrating environmental, economic, and social goals.

Empowerment: From Awareness to Action

Raising Awareness

Awareness of climate change and its impacts is the first step toward meaningful action. However, awareness alone is not enough; it must be coupled with a sense of agency and the capacity to act.

Education and Advocacy

Climate Education: Integrating climate education into school curricula and community programs fosters a deep understanding of climate issues and empowers individuals to take informed action.

Public Campaigns: Advocacy and public awareness campaigns raise the profile of climate issues and mobilize public support for policy changes and sustainable practices.

Media and Storytelling

Narratives and Media: Media, literature, and art play a critical role in shaping public perception and understanding of climate change. Effective storytelling can inspire action and foster a sense of connection to the natural world.

Social Media Activism: Social media platforms amplify voices, spread awareness, and mobilize grassroots movements, making climate activism more accessible and widespread.

Building Capacity

Empowerment involves building the capacity of individuals and communities to implement sustainable practices and advocate for systemic change.

Community Engagement

Local Initiatives: Community-based initiatives, such as urban gardening, local energy projects, and conservation efforts, demonstrate the power of collective action and local leadership.

Participatory Governance: Involving communities in decision-making processes ensures that policies and practices reflect local needs and knowledge, enhancing their effectiveness and acceptance.

Policy and Institutional Support

Policy Frameworks: Strong policy frameworks that support renewable energy, conservation, and sustainable development provide the necessary infrastructure for individual and collective action.

Institutional Partnerships: Partnerships between governments, businesses, NGOs, and educational institutions foster collaborative efforts and leverage resources for greater impact.

Individual Actions

Lifestyle Choices: Individuals can reduce their carbon footprint through choices such as adopting plant-based diets, reducing waste, and using sustainable transportation.

Civic Engagement: Voting for leaders who prioritize climate action, participating in climate protests, and supporting climate-friendly businesses are ways individuals can contribute to systemic change.

Case Studies

Germany: Germany's Energiewende (energy transition) showcases the power of policy support and community engagement in driving the shift to renewable energy.

Indigenous Communities in the Amazon: Indigenous communities' efforts to protect their lands and preserve traditional knowledge highlight the importance of local leadership and cultural resilience.

Future Directions: Continuing the Journey

Innovative Solutions

The journey toward a sustainable future requires continuous innovation in technology, policy, and social practices.

Technological Advancements

Clean Energy: Ongoing research and development in renewable energy technologies, energy storage, and smart grids will further reduce carbon emissions and increase energy resilience.

Climate Mitigation Technologies: Innovations in carbon capture, sustainable agriculture, and green infrastructure will enhance our capacity to mitigate and adapt to climate impacts.

Policy Innovations

Climate Policy: Developing and implementing robust climate policies at local, national, and international levels will drive systemic change and ensure accountability.

Economic Incentives: Policies that incentivize sustainable practices, such as carbon pricing, subsidies for renewable energy, and penalties for pollution, will align economic activities with environmental goals.

Strengthening Global Cooperation

Climate change is a global challenge that requires global solutions. Strengthening international cooperation and solidarity is essential for effective climate action.

International Agreements

Paris Agreement: Continued commitment to the Paris Agreement and its goals is crucial for coordinated global efforts to limit global warming.

Biodiversity and Conservation: International agreements on biodiversity and conservation, such as the Convention on Biological Diversity, play a vital role in protecting ecosystems and species.

Global Solidarity

Climate Justice: Ensuring that climate action addresses social and economic inequalities and supports vulnerable populations is fundamental to achieving climate justice.

Shared Knowledge and Resources: Sharing knowledge, technologies, and resources across borders enhances global capacity to tackle climate change and build resilience.

Personal and Collective Commitment

The journey toward a sustainable future is ongoing and requires sustained personal and collective commitment.

Continuous Learning

Lifelong Learning: Engaging in lifelong learning about climate science, sustainability practices, and policy developments empowers individuals to stay informed and proactive.

Adapting to Change: Embracing change and continuously adapting personal and collective practices to emerging challenges and opportunities is essential for resilience.

Building Hope

Positive Visioning: Fostering positive visions of the future, where sustainable practices, technological innovations, and social equity are realized, inspires hope and motivates action.

Celebrating Successes: Recognizing and celebrating successes in climate action, no matter how small, reinforces the importance of collective effort and progress.

Case Studies

Scotland: Scotland's ambitious climate targets and commitment to a just transition demonstrate the power of visionary leadership and policy innovation.

Kenya: Kenya's investment in renewable energy, particularly geothermal power, highlights the potential for technological innovation and international cooperation in achieving sustainability goals.

This concluding chapter synthesizes the knowledge and insights gained throughout the book, emphasizing the interconnected nature of climate change and human identity. It highlights the importance of moving from awareness to action, building capacity, and fostering global cooperation. As we continue the journey toward a sustainable future, innovation, resilience, and collective commitment will be essential. By integrating knowledge and action, we can create a world where both the environment and human identity thrive in harmony.

Epilogue: A Call to Action

Individual Responsibility: Everyday Choices

The Power of Personal Action

Individual responsibility plays a crucial role in combating climate change. Each person's choices, no matter how small, can collectively lead to significant impacts. The epilogue emphasizes the importance of recognizing the power of personal action in shaping a sustainable future.

Sustainable Living Practices

Reducing Consumption: Adopt minimalistic lifestyles that focus on reducing, reusing, and recycling. This can significantly cut down waste and lower your carbon footprint.

Energy Efficiency: Utilize energy-efficient appliances, switch to LED lighting, and practice mindful energy use at home. Small changes like unplugging devices when not in use and optimizing heating and cooling systems contribute to energy conservation.

Sustainable Transportation: Opt for public transportation, carpooling, cycling, or walking instead of driving alone. If feasible, transition to electric or hybrid vehicles to reduce greenhouse gas emissions.

Responsible Consumption

Eco-Friendly Products: Choose products with minimal packaging, made from sustainable materials, and those produced by environmentally responsible companies.

Plant-Based Diet: Reduce meat consumption and incorporate more plant-based foods into your diet. Animal agriculture is a significant contributor to greenhouse gas emissions, and shifting towards plant-based diets can lessen this impact.

Local and Seasonal Foods: Support local farmers and eat seasonal produce to reduce the carbon footprint associated with food transportation and storage.

Personal Advocacy

Education and Awareness: Continuously educate yourself and others about climate change and sustainable practices. Knowledge empowers informed decision-making and fosters a culture of environmental stewardship.

Social Media Engagement: Use social media platforms to spread awareness, share information, and mobilize others. Online activism can reach a wide audience and inspire collective action.

Political Participation: Vote for leaders and policies that prioritize climate action. Engage in local government meetings, advocate for climate-friendly policies, and support initiatives that promote sustainability.

Case Studies

Greta Thunberg: Greta's individual initiative of striking for the climate has inspired a global movement, highlighting the impact one person can have.

The Zero Waste Family: The Johnson family's journey towards a zero-waste lifestyle demonstrates the feasibility and benefits of reducing personal waste and consumption.

Collective Effort: Community and Global Initiatives

The Strength of Community

Communities have the power to drive significant change through collective effort. By working together, communities can implement local solutions that contribute to global sustainability goals.

Local Initiatives

Community Gardens: Establishing community gardens promotes local food production, reduces food miles, and fosters a sense of community. It also educates participants about sustainable agriculture.

Renewable Energy Projects: Community-driven renewable energy projects, such as solar cooperatives, can provide clean energy, reduce reliance on fossil fuels, and create local jobs.

Waste Management Programs: Local recycling programs, composting initiatives, and waste reduction campaigns help communities manage waste sustainably and reduce landfill use.

Building Resilience

Disaster Preparedness: Community-based disaster preparedness plans enhance resilience to climate impacts such as extreme weather events. Collaborative planning ensures that resources and support are available to those most in need.

Green Infrastructure: Investing in green infrastructure, such as parks, green roofs, and urban forests, enhances urban resilience, improves air quality, and provides recreational spaces.

Global Solidarity

International Cooperation: Climate change is a global issue that requires global solutions. International agreements, such as the Paris Agreement, foster cooperation and commitment to reducing emissions and mitigating climate impacts.

Knowledge Sharing: Sharing best practices, technologies, and research across borders enhances global capacity to address climate change. Collaborative research and development can lead to innovative solutions.

Climate Justice: Ensuring that climate action addresses social and economic inequalities is crucial. Supporting vulnerable communities, both locally and globally, fosters a fair and just transition to a sustainable future.

Case Studies

Transition Towns Movement: The Transition Towns movement empowers communities to become more sustainable and resilient through local action and collaboration.

Global Climate Strikes: The global climate strike movement, initiated by youth activists, has mobilized millions worldwide, demonstrating the power of collective action.

Sustaining Change: A Unified Future

Long-Term Commitment

Sustaining change requires a long-term commitment to continuous improvement and adaptation. It involves embedding sustainability into the fabric of everyday life, policy, and governance.

Institutional Support

Government Policies: Strong, consistent policies at local, national, and international levels are essential for sustaining climate action. These policies should support renewable energy, conservation, sustainable agriculture, and climate resilience.

Corporate Responsibility: Businesses must adopt sustainable practices, reduce their carbon footprint, and contribute to climate solutions. Corporate responsibility includes transparent reporting, green innovations, and ethical supply chains.

Education and Awareness

Lifelong Learning: Climate education should not be limited to schools. Lifelong learning opportunities, such as community workshops, online courses, and public lectures, keep individuals informed and engaged.

Public Awareness Campaigns: Ongoing public awareness campaigns can reinforce the importance of sustainable practices and keep climate change at the forefront of public consciousness.

Visionary Leadership

Inspirational Leaders: Visionary leaders, whether in politics, business, or civil society, can inspire and mobilize others. Their commitment to sustainability and ethical governance sets an example and drives progress.

Youth Leadership: Encouraging and supporting youth leadership is crucial. Young people bring energy, innovation, and a fresh perspective to climate action.

Hope and Optimism

Positive Narratives: Promoting positive narratives about the future and celebrating successes, no matter how small, fosters hope and motivation. Highlighting success stories and innovations shows that change is possible.

Collective Vision: A shared vision for a sustainable future unites individuals and communities. This vision should be inclusive, equitable, and focused on the well-being of both people and the planet.

Case Studies

New Zealand: New Zealand's comprehensive approach to sustainability, including policies on renewable energy, conservation, and social equity, exemplifies long-term commitment and visionary leadership.

Bhutan: Bhutan's commitment to gross national happiness, which includes environmental conservation and sustainable development, demonstrates the integration of well-being and sustainability.

This epilogue emphasizes the critical role of both individual and collective action in addressing climate change. By making conscious everyday choices, participating in community initiatives, and supporting global efforts, each person can contribute to a unified future where sustainability and resilience are at the forefront. Sustaining change requires long-term commitment, visionary leadership, and a shared vision of hope and optimism. Together, we can create a world that thrives in harmony with the environment, ensuring a sustainable future for generations to come.

Thank You for Reading

Thank you for choosing "Climate Change and Its Effects on Human Identity." We hope this book has provided you with valuable insights into the profound connections between our changing climate and human identity. Your feedback is incredibly important to us and helps guide future readers.

If you found this book informative and engaging, please consider leaving a positive review on Amazon. Your support is greatly appreciated and contributes to spreading awareness about the intricate and critical relationships between climate change and human identity. Thank you for being part of this journey towards a deeper understanding and a more resilient, sustainable future.

Warm regards,

FAISAL JAMIL

For Your Feedback and Reviews!

http://www.amazon.com/author/faisal.jamil

Email: faisaljamilauthor@gmail.com

www.ingramcontent.com/pod-product-compliance
Lightning Source LLC
Chambersburg PA
CBHW071921210526
45479CB00002B/502